Durability of Carbon Fiber Reinforced Plastics

Koyanagi presents a concise and practical guide to using a micromechanics approach to predict the strength and durability of unidirectionally aligned continuum carbon-fiber-reinforced plastics (CFRPs).

As the use of composite materials becomes more widespread in various fields, material durability is becoming an increasingly important consideration, particularly with regard to UN Sustainable Development Goals. Using more durable composite materials would help with achieving these goals. Because the failure of composite materials proceeds via the accumulation of microfailures and micro damage, a micromechanics approach is indispensable for estimating precise durability.

In this practical guide, Koyanagi describes this approach and explains the precise durability of the composite materials with regard to the time dependence of microfailures. This book first explains the strength and durability of unidirectionally aligned continuum CFRPs. It then individually addresses fiber, resin, and the interface between the two on the basis of their micromechanics and introduces these components' time and temperature dependences. Koyanagi uses finite element analysis and theoretical models to integrate the characteristics of the three components to explain the macro properties of the CFRPs. Various characteristics regarding strength and durability of CFRPs are also presented.

This book is a valuable resource for researchers in academia and industry who work with composite materials. It will enable them to design composite structures, ensure their durability, evaluate them, and develop more durable composite materials.

Jun Koyanagi is Professor at Tokyo University of Science (TUS), Katsushika, Tokyo, Japan, and leader of the Composite Materials Laboratory there. He gained his PhD from Waseda University Tokyo, Japan, in 2006. Before joining TUS, he was with the Institute of Space and Astronautical Science, Japan Aerospace Exploration Agency (JAXA).

Durability of Carbon Fiber Reinforced Plastics

Jun Koyanagi

CRC Press
Taylor & Francis Group
Boca Raton London New York

CRC Press is an imprint of the
Taylor & Francis Group, an **informa** business

First edition published 2024
by CRC Press
4 Park Square, Milton Park, Abingdon, Oxon, OX14 4RN

and by CRC Press
2385 NW Executive Center Drive, Suite 320, Boca Raton FL 33431

© 2024 Jun Koyanagi

CRC Press is an imprint of Informa UK Limited

ISBN: 978-1-032-44233-4 (hbk)
ISBN: 978-1-032-44234-1 (pbk)
ISBN: 978-1-003-37113-7 (ebk)

DOI: 10.1201/9781003371137

Typeset in Times LT Std
by Apex CoVantage, LLC

Contents

Introduction

1.1 SIGNIFICANCE OF MICROMECHANICS

Continuous carbon-fiber-reinforced composites (CFRPs) exhibit exceptional specific strength and specific stiffness, rendering them invaluable for a wide range of applications. Despite CFRP's extensive adoption spanning more than half a century, the quest for strength and durability remains paramount. Within this context, the author perceives that the crux of this matter lies in the prognostication grounded in the domain of micromechanics.

The failure mechanism of CFRP encompasses the progressive accumulation of fiber failure, matrix failure, and interface debonding between the fiber and matrix. It is of utmost significance to scrutinize the microscale failure phenomena from a microscopic standpoint, focusing on the localized stress states surrounding specific fiber, interface, and matrix failures. To this end, finite element analysis emerges as an indispensable tool. This book endeavors to elucidate the predictive and evaluative capabilities of CFRP strength and durability through comprehensive utilization of numerical simulations, primarily rooted in finite element analysis methodologies.

Achieving accurate predictions and assessments of CFRP durability necessitates a thorough understanding of microscale failure mechanisms. While fibers and interfaces can generally be considered independent of external factors like time, temperature, and loading history, the behavior of the matrix is more intricate. As a polymer material with viscoelastic properties, the matrix undergoes failure that is influenced by time, temperature, and loading history. Recognizing and comprehending these factors are vital for providing a preliminary estimation of CFRP durability. However, for precise and reliable predictions, additional considerations are necessary. This book introduces methodologies that employ numerical simulations to forecast CFRP durability, encompassing these factors and their intricate interplay.

The author has contributed numerous articles that explore the durability of composite materials from a micromechanics perspective [1–11]. These studies aim to challenge the validity of conventional evaluation methodologies for

DOI: 10.1201/9781003371137-1

composite material durability, which often overlook the role of micromechanics, relying primarily on experimental investigations [12–15]. A central theme in these articles revolves around the application of the time–temperature superposition principle. This principle allows for the acceleration of time at elevated temperatures. While the time–temperature superposition principle is commonly applied to "time-dependent deformation" in polymer materials, its applicability to time-dependent failure of polymers, and even more so to composite materials, remains a subject of debate that necessitates a thorough examination from a micromechanics standpoint. The book takes a cautious approach and strictly applies the time–temperature superposition principle solely to polymer deformation, avoiding unwarranted assumptions that have been observed in conventional studies. In the following chapters, this book delves into the durability of composite materials based on micromechanics.

1.2 OVERVIEW OF THIS BOOK

Chapter 2 delves into the mechanical properties of the fiber–matrix interface. Since the interface is planar, it experiences only one normal stress and two shear stresses. These two shear stresses can be combined into a single component by taking their square average, resulting in a two-component stress state at the interface: normal stress and shear stress. The strength of the interface is assessed using a failure envelope plotted on the normal stress–shear stress plane. Based on extensive experimental investigations conducted by the author, a parabolic criterion is recommended as the appropriate failure envelop. Furthermore, it is concluded that the interface strength is not influenced by time or temperature. This chapter presents the methodology employed to reach this conclusion, highlighting versatile experiments and corresponding finite element analyses that were conducted. Additionally, this chapter discusses the application of molecular dynamics simulations to investigate the interface properties. It is reasonable to expect that the molecular structure at the interface can provide insights into predicting the interface properties. Several potential approaches in this regard are introduced and explored within the chapter. By utilizing molecular dynamics simulations, it becomes possible to gain a deeper understanding of the interface and its associated properties.

Chapter 3 delves into the constitutive relationship of the matrix resin, which is a polymer material. The chapter begins by presenting the linear viscoelastic constitutive equation, followed by the inclusion of the nonlinear counterpart. Subsequently, the time–temperature superposition principle is introduced, and, finally, the consideration of damage is addressed. In the case of polymer materials, predicting residual strength after one or multiple usages is challenging. However, the utilization of entropy as a means to define failure proves to be highly advantageous. While the concept of entropy-based damage law has gained prominence in the realm of metals since around 2010, its application to polymer materials is a pioneering effort undertaken by the authors. Entropy, defined as the ratio of dissipated energy to temperature, serves as a convenient failure criterion. Only those well-versed in inelastic deformation phenomena, such as viscoelastic and viscoplastic behaviors, can effectively employ entropy as an index of failure. Hence, a strong foundation in viscoelasticity plays a crucial role in this context.

Chapter 4 introduces various models for estimating the strength of composites based on the strengths of their constituent components: fibers, matrix, and interface. Specifically, the chapter addresses the prediction of fiber-axial tensile strength, fiber-axial compressive strength, fiber-axial shear strength, and fiber-transverse strengths. Among these, predicting the fiber-axial tensile strength and fiber-axial compressive strength through finite element analysis proves challenging due to the involvement of fiber failures. The high aspect ratio of fibers makes it difficult to create realistic finite element models. To address this, the author proposes the SFF (Simultaneous Fiber Failure) model and the widely recognized microbuckling model as means to estimate these two strengths. On the other hand, the remaining strengths can be estimated using finite element models that incorporate periodic unit cell simulations with cohesive zone modeling and continuum damage mechanics. Each of these models is individually explained in this chapter. Potentially, these models are extended to estimate the long-term durability of composite materials by considering the time-dependent failure of the resin matrix. It is important to note that the fiber strength and interface strength are not influenced by time or temperature variations.

Chapter 5 of the book focuses on comparing the strengths and durability obtained from experiments with those obtained from micromechanical analyses, specifically examining cases such as fiber-transverse tensile strength and fiber-axial tensile strength. Based on the characteristics of matrix strength and interface strength discussed earlier, the failure mode transitions from matrix failure dominance to interface failure dominance as the strain rate increases in transverse tensile failure. This is due to the time-independent nature of the interface strength and the time-dependent behavior of the matrix strength. The simulated results and experimental results exhibit good quantitative agreement. Notably, the chapter highlights the reasonable estimation of residual strength after a certain duration of loading using an entropy-based damage criterion. The simulation methodology is described in detail, providing insights into the numerical analysis used. Additionally, the chapter introduces the time and temperature dependence of fiber-axial tensile strength. In this estimation, the SFF model is extensively employed, allowing for the reasonable estimation of long-term durability for various CFRPs, even those exhibiting complex experimental results.

Chapter 6 takes a step beyond micromechanics and delves into "mesoscale" numerical simulations. The primary focus of this chapter is on the damage incurred by CFRP plies under cyclic loading conditions. As depicted in Figure 1.1, a comprehensive understanding of composite material durability requires insights into molecular-level damage, microscale failure mechanics, and mesoscale damage accumulation. While the chapter does not explicitly address the establishment of a quantitative link between microscale and mesoscale simulations, it introduces a novel finite element approach for simulating strength reduction resulting from entropy increase, which is closely related to dissipated energy. This pioneering concept, though challenging in terms of parameter identification, holds significant utility. With this element, it becomes possible to predict not only degradation and damage states but also remaining lifetime and residual strength, even when subjected to varying stress amplitudes, stress levels, and frequencies in cyclic loading scenarios. Such an element has the potential to catalyze a paradigm shift in the study of composite material durability, which has predominantly followed a monotonic trajectory since the 1980s.

Chapter 7 presents an exploration of future perspectives in the field. As our understanding of composite material durability continues to evolve, several key areas of focus and potential avenues for research emerge.

First, there is a need to develop predictive models for interlayer delamination in CFRP laminates. By accurately assessing interlayer debonding, we can enhance the overall durability of composite structures.

Second, the evaluation of durability in discontinuous fiber CFRP composites deserves attention. As the use of these materials becomes more prevalent, it is crucial to understand their unique failure mechanisms and develop appropriate testing methodologies.

Third, the investigation of durability under random loading conditions is essential. Many real-world applications involve complex and unpredictable loading patterns, necessitating a comprehensive understanding of how composites behave under such conditions.

Additionally, the development of simplified models that can provide quick estimations of composite strength and durability is of practical importance. These models can serve as valuable tools in the design and optimization of composite structures.

Lastly, the concept of design from the molecular level holds promise for advancing the field. By harnessing molecular dynamics simulations, we can gain valuable insights into the relationship between molecular structure and composite performance, paving the way for more tailored and optimized composite materials.

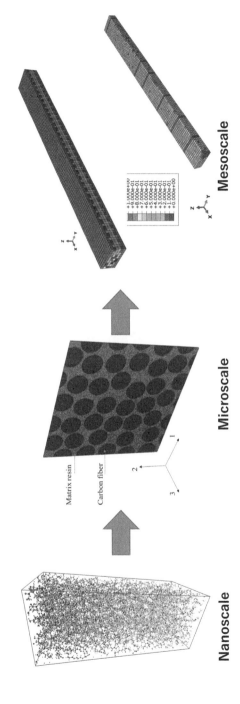

Nanoscale   **Microscale**   **Mesoscale**

FIGURE 1.1   Multiscale simulations from molecular level to mesoscale for elucidating composite durability.

It is worth noting that while this book contributes to the understanding and prediction of CFRP durability, further research is still needed to address challenges such as more efficient computational frameworks and the impact of heat generation during cyclic loading. These endeavors will be the subject of future studies in this field.

REFERENCES

[1] Jun Koyanagi, Asa Mochizuki, Ryo Higuchi, V.B.C. Tan, T.E. Tay, Finite element model for simulating entropy-based strength-degradation of carbon-fiber-reinforced plastics subjected to cyclic loadings, International Journal of Fatigue 165 (2022) 107204.

[2] J. Koyanagi, Comparison of a viscoelastic frictional interface theory with a kinetic crack growth theory in unidirectional composites, Composites Science and Technology 69(13) (2009) 2158–2162.

[3] J. Koyanagi, Durability of filament-wound composite flywheel rotors, Mechanics of Time-Dependent Materials 16(1) (2012) 71–83.

[4] J. Koyanagi, H. Hatta, F. Ogawa, H. Kawada, Time-dependent reduction of tensile strength caused by interfacial degradation under constant strain duration in UD-CFRP, Journal of Composite Materials 41(25) (2007) 3007–3026.

[5] J. Koyanagi, G. Kiyota, T. Kamiya, H. Kawada, Prediction of creep rupture in unidirectional composite: Creep rupture model with interfacial debonding and its propagation, Advanced Composite Materials: The Official Journal of the Japan Society of Composite Materials 13(3–4) (2004) 199–213.

[6] J. Koyanagi, M. Nakada, Y. Miyano, Prediction of long-term durability of unidirectional CFRP, Journal of Reinforced Plastics and Composites 30(15) (2011) 1305–1313.

[7] J. Koyanagi, M. Nakada, Y. Miyano, Tensile strength at elevated temperature and its applicability as an accelerated testing methodology for unidirectional composites, Mechanics of Time-Dependent Materials 16(1) (2012) 19–30.

[8] J. Koyanagi, Y. Sato, T. Sasayama, T. Okabe, S. Yoneyama, Numerical simulation of strain-rate dependent transition of transverse tensile failure mode in fiber-reinforced composites, Composites Part A: Applied Science and Manufacturing 56 (2014) 136–142.

[9] J. Koyanagi, S. Yoneyama, A. Nemoto, J.D.D. Melo, Time and temperature dependence of carbon/epoxy interface strength, Composites Science and Technology 70(9) (2010) 1395–1400.

[10] J. Koyanagi, A. Yoshimura, H. Kawada, Y. Aoki, A numerical simulation of time-dependent interface failure under shear and compressive loads in single-fiber composites, Applied Composite Materials 17(1) (2009) 31–41.

[11] J. Koyanagi, A. Yoshimura, H. Kawada, Y. Aoki, A numerical simulation of time-dependent interface failure under shear and compressive loads in single-fiber composites, Applied Composite Materials 17(1) (2010) 31–41.

[12] Y. Miyano, M. Nakada, H. Cai, Formulation of long-term creep and fatigue strengths of polymer composites based on accelerated testing methodology, Journal of Composite Materials 42(18) (2008) 1897–1919.

[13] Y. Miyano, M. Nakada, H. Kudoh, R. Muki, Prediction of tensile fatigue life for unidirectional CFRP, Journal of Composite Materials 34(7) (2000) 538–550.

[14] Y. Miyano, M. Nakada, R. Muki, Applicability of fatigue life prediction method to polymer composites, Mechanics Time-Dependent Materials 3(2) (1999) 141–157.

[15] M. Nakada, Y. Miyano, Advanced accelerated testing methodology for long-term life prediction of CFRP laminates, Journal of Composite Materials 49(2) (2015) 163–175.

Mechanical Properties of Interface Between Fiber and Matrix

2

2.1 INTRODUCTION

In light of the growing prominence of polymer matrix composites and their diverse applications, such as serving as the primary structural material for cutting-edge air vehicles, understanding the intricacies of interface strength becomes crucial. This strength, which is the adhesion between the reinforcing fiber and the polymer matrix, remains a key challenge in the development of composite materials. The mechanical properties of composites are influenced by the components of the fiber, the matrix, and the interface itself. It is important to note that high adhesive strength is not always beneficial, as it can sometimes lead to reduced composite performance [1]. Consequently, the accurate measurement of interface mechanical properties remains crucial.

Interface failure typically occurs in normal and shear modes, often as a mixed-mode combination. To effectively evaluate interface strength, both normal and shear strengths must be assessed, and the failure envelope under a combined stress state must be considered, as shown in Figure 2.1 [2]. In this figure, the x-axis represents the normalized normal stress, calculated as the normal stress divided by the interface strength in pure normal mode. Similarly, the y-axis displays the normalized shear stress, derived from

DOI: 10.1201/9781003371137-2

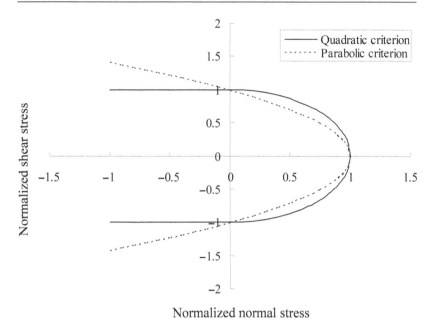

FIGURE 2.1   Interface failure envelopes in normalized normal stress–normalized shear stress plane.

the shear stress divided by the interface strength in pure shear mode. This figure allows for the construction of an interface failure envelope. When normal stress is positive, the interface strength is defined as a function of both shear and normal stresses. However, when normal stress is negative, indicating the presence of compressive stress on the interface, it becomes debatable whether the compressive stress influences the interface failure envelope. This uncertainty arises because the interface does not fail solely due to compressive stress. Consequently, two types of failure envelopes are introduced: the quadratic criterion and the parabolic criterion. The quadratic criterion posits that compressive stress does not impact interface failure, and only tensile stress plays a role. A general ellipsoid law is applied when tensile and shear stresses coexist. Conversely, the parabolic criterion assumes a parabolic relationship for the interface failure criterion. The two criteria are expressed in

$$\left(\frac{\langle t_n \rangle}{Y_n}\right)^2 + \left(\frac{t_s}{Y_s}\right)^2 = 1 \qquad\qquad \text{Eq. (2.1)}$$

$$\frac{t_n}{Y_n} + \left(\frac{t_s}{Y_s}\right)^2 = 1 \qquad\qquad \text{Eq. (2.2)}$$

Here Y is strength, t is stress, <> is McCaulay bracket, subscript n means normal direction, and subscript s means shear direction. As we assess interface strength, it is vital to consider whether we are evaluating the normal or shear strength of the interface. By the end of this chapter, we will establish a quantitative relationship between normal and shear strengths, drawing from experimental results.

Another crucial aspect to consider when evaluating interface strength is "stress distribution". Interface strength is often defined as the experimental force required for interface failure divided by the area of the failed interface. This approach inherently assumes a uniform stress distribution across the entire failed interface, without considering potential inaccuracies. Such an assumption may only be reasonably applied when dealing with perfectly plastic materials, which might be a rare occurrence, particularly when working with polymer resins. Therefore, it is essential to account for stress distribution and focus on the stress value where the interface starts failing. With these points in mind, we proceed to the following sections.

2.2 HOW TO EVALUATE INTERFACE STRENGTH

In this section, we will discuss various methodologies that have been developed to evaluate interfacial strength by utilizing single fibers and matrix resins. As is well-known, carbon-fiber-reinforced polymers (CFRPs) are composed of a multitude of fibers embedded within a surrounding matrix, rendering the extraction of interface strength from bulk composite specimens a complex endeavor. As a result, concentrating on the interface between an individual fiber and the matrix resin proves to be a more viable strategy for assessing interface mechanical strength. A diverse array of evaluation methods have been employed in the scientific literature, including the single-fiber fragmentation test [3, 4], single-fiber pull-out test [5], cruciform specimen test [6, 7], and microbond test [8, 9]. In the ensuing sections, we will provide a comprehensive examination of some of these methodologies, elaborating on their underlying principles and experimental protocols.

2.2.1 Microbond Test [8, 9]

The test is that microbond attached with a single fiber is debonded in order to evaluate interface debonding strength. The specimen consists of carbon fiber and matrix resin. The droplet is removed by this equipment, and the load–displacement relationship is recorded. If the interface exhibits high strength, a considerable load will be required to remove the droplet. However, if the force needed to induce interface failure exceeds the fiber's inherent strength, the test becomes ineffective. In such scenarios, the fiber experiences failure without any corresponding interface failure. Consequently, it becomes impossible to evaluate the interface strength under these circumstances.

Following the experiment, it is necessary to conduct finite element analysis to evaluate the stress distribution. Some researchers calculate the interface strength by determining the shear stress value, which is the load divided by the failure interface area [1]. This method, referred to as "apparent interface strength", assumes that the interface stress is uniformly applied across the entire interface. However, in reality, stress distribution is nonuniform, and it is essential to consider the point at which the interface begins to fail. The stress value at this failure initiation point may be considered the true interface strength. To evaluate the true interface strength, numerical simulations must be conducted under experimental conditions [8, 9]. In these simulations, the nonlinearity of the constitutive equation for the matrix resin plays a crucial role.

By combining the microbond test with corresponding numerical simulations, the interface strength can be accurately evaluated, addressing issues such as size dependence of interface strength. It is important to note that the true interface strength is consistently higher than the apparent interface strength. While the apparent interface strength can be useful for qualitative comparisons, it is crucial to bear in mind that its value is likely an underestimation of the actual interface strength.

2.2.2 Single-Fiber Pull-Out Test Using Pin-Holed Plate

The innovative equipment, "single fiber pull-out test using pin-holed plate", is illustrated in Figure 2.2. The fundamental concept remains the same as the conventional single-fiber pull-out test. The primary distinction from the traditional method is the presence of a small pocket that is filled with the matrix. This apparatus allows for controlled embedding of the fiber, thereby enabling

FIGURE 2.2 Experimental equipment for single-fiber pull-out test using pin-holed plate.

the management of the pulled-out fiber length. As a result, even interfaces with very high strength can be examined without fiber failure, as seen in the microbond test.

Figure 2.3 presents a magnified view of the area surrounding the fiber and pin-holed plate filled with resin, both before the experiment (left) and after the experiment with a defined pulled-out length (right). Before the experiment, the center of the pin hole and the fiber location coincide. After the fiber is pulled out, the fiber shifts slightly, as depicted in the right-hand photograph. By examining the specimen after the experiment, the pulled-out length can be determined through the presence of the resin meniscus shown in the right-hand photograph. This equipment was developed by Shin-so-sha Inc. Based on the author's experience, this apparatus is considered to be the most effective for evaluating interface strength.

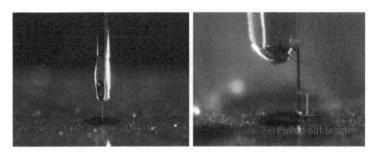

FIGURE 2.3 Magnified view around fiber and pin-holed plate filled with resin: before experiment (left) and after experiment with definition of pulled-out length (right).

Similar to the microbond test, the single-fiber pull-out test also necessitates performing numerical simulations to evaluate accurate interface stress distribution. By implementing finite element analysis under experimental conditions equivalent to those of the test, the true interface strength can be obtained.

2.2.3 Cruciform Specimen Test [5, 6]

It is important to note that, initially, this section focuses solely on the interface between glass/epoxy materials. The specimen geometry and dimensions for the cruciform specimen test are illustrated in Figure 2.4. A single fiber is embedded at the same angle as the arm's direction in the cruciform specimen. This configuration eliminates interface debonding, resulting from stress singularity at the specimen's edge. The interface debonding occurs within the specimen's internal part, allowing for a more accurate evaluation of interface strength. Naturally, numerical simulation is necessary to determine the accurate interface stress at the point of failure.

It should be emphasized that this method provides interface "normal strength" when the cruciform arms are perpendicular to the loading direction, as opposed to the "shear strength" discussed in previous sections. Additionally, by altering the angle of the cruciform arms, as shown in Figure 2.4, the stress ratio of mixed-mode combined stress state at interface failure can be controlled. Consequently, the cruciform specimen method enables a discussion of interface failure envelopes. Detailed results can be found in the authors' article [6]. Key findings indicate that the pure shear interface strength is approximately 1.3 to 1.4 times greater than the pure normal interface strength. This

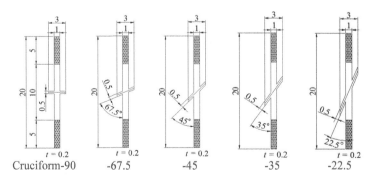

FIGURE 2.4 Specimen geometry and dimensions.

conclusion aligns with the understanding that interface shear toughness is greater than interface normal toughness [10]. While it is not always the case that higher toughness materials possess higher strength, this correlation may be applicable to interface properties since an ideal interface is planar, without volume, and undefined in terms of deformation—only surface energy can be defined. Hence, a higher toughness interface may exhibit higher strength. These results are summarized in Figure 2.5.

2.2.4 Interface Failure Envelope

Figure 2.5 depicts the interface failure envelope, similar to Figure 2.1. These are experimental results obtained through a combination of experiments and numerical simulations to determine accurate interface strength. The experiments include the cruciform specimen test and the single-fiber pull-out test, which uses the conventional method without a pin-hole plate. It is important to note that both normal and shear stresses occur at the interface failure

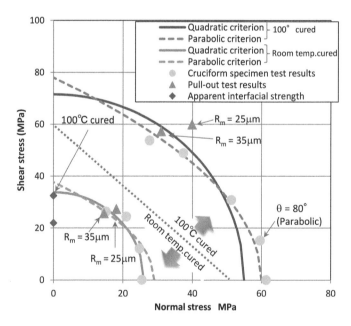

FIGURE 2.5 Interface failure envelope for two kinds of glass/epoxy interface; one has relatively strong interface bonding and another has relatively weak one, obtained from cruciform specimen test and single-fiber pull-out test and approximated curves for the results based on quadratic criterion and parabolic criterion.

initiation location in the single-fiber pull-out test, and these stresses are calculated using FEA. In Figure 2.5, two types of interfaces are introduced: one with relatively higher interface strength and another with relatively lower interface strength, fabricated under 100°C curing conditions and room temperature curing conditions, respectively. Solid symbols represent experimental results, while approximated curves based on quadratic and parabolic criteria are also depicted.

The following points are clarified by the results:

I. The results obtained from the cruciform specimen test and the pull-out test are in good agreement with each other. Similar values are derived from different methods, which is reasonable since the interface itself is identical. This indicates that the evaluation method, consisting of experiment and numerical simulation, is quantitatively valid.

II. The pull-out test result should be considered as a combined stress state, that is, normal and shear stresses occur at the interface failure initiation point. It is not reasonable to assume that this test solely provides interface shear strength.

III. The interface strength in pure normal mode is less than that in pure shear mode. The pure shear strength is 1.2 to 1.4 times greater than the pure normal strength.

IV. It is not clear which criterion, quadratic or parabolic, is more suitable for the interface failure envelope, as both criteria produce similar curves within the range of these results.

2.3 TIME AND TEMPERATURE DEPENDENCE OF INTERFACE STRENGTH

In this section, we explore the time and temperature dependence of interface strength, which is crucial when discussing the long-term durability of composite materials. The degradation of interface mechanical properties over time is a key factor in ensuring the long-term durability of these materials. Additionally, many researchers employ accelerated testing methods to investigate the effects of time and temperature on composite materials' performance.

Accelerated testing methods involve subjecting materials to higher temperatures, increased loads, or more aggressive environmental conditions than they would typically experience during their expected service life. The objective is to induce failure or degradation more quickly, allowing researchers to draw conclusions about the material's long-term performance based on shorter experimental time frames.

It is essential to understand the relationship between interface strength and both time and temperature, as these factors can influence the material's overall performance and durability. For instance, an exposure to elevated temperatures may weaken the interface over time, reducing the composite material's load-bearing capacity. Similarly, long-term exposure to environmental factors, such as moisture, may also degrade the interface and compromise the composite's integrity.

By investigating the time and temperature dependence of interface strength, researchers can develop a better understanding of composite materials' long-term performance and durability [11, 12]. This knowledge can inform material selection, design, and maintenance strategies, ultimately improving the reliability and lifespan of composite materials in various applications.

2.3.1　Time and Temperature Dependence of Interface "Normal" Strength

Three articles have been published concerning the time and temperature dependence of interface normal strength [7, 13, 14]. Here, we present some of the key findings from these studies. The temperature dependence of normal strength for an interface composed of glass fiber and epoxy resin was investigated using the cruciform specimen method with temperature-controlled equipment [7]. It is important to note that numerical analysis was also employed to determine the stress value at the interface failure initiation point. Within the range of the conducted experiments, the interface normal strength appears to be independent of temperature [7]. This finding suggests that the interface's normal strength remains relatively stable across the tested temperature range, which is an important consideration for the design and application of composite materials in various environments.

The second article discusses the strain-rate dependence of interface normal strength, which is equivalent to its time dependence [13]. In this study, the interface normal strength was examined using transverse tensile tests of unidirectional carbon-fiber-reinforced composite materials. The findings suggest that the interface normal strength in CFRP is not time-dependent. The third paper is similar to the second one, exploring both the time and temperature dependence of interface strength in CFRP. This was also done using transverse tensile tests of unidirectional CFRP under various tensile speeds and temperatures. As illustrated in Figure 2.6, the failure mode can be classified into two categories: interface failure dominant mode and matrix failure dominant mode.

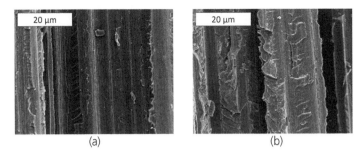

(a)   (b)

FIGURE 2.6 Failure surfaces of transverse tensile tests of unidirectional CFRP; (a) interface dominant mode is seen when the test conditions are relatively low temperature and high strain rate, on the other hand, (b) matrix failure dominant model is seen with opposite test conditions.

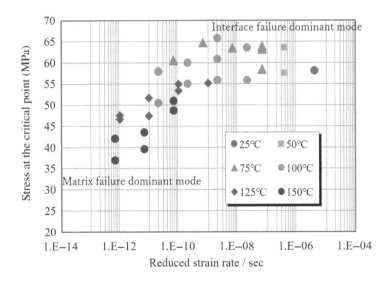

FIGURE 2.7 Relationship between the strengths of interface or matrix and reduced strain rate in which time–temperature superposition principle is considered; the failure mode is classified into matrix failure and interface failure, and interface failure is not time and temperature dependent.

The former is observed in test results obtained under relatively low temperatures and high strain rates, while the latter occurs under opposite conditions. The ultimate conclusion of this study is depicted in Figure 2.7. In this figure, the time–temperature superposition principle is applied, with the x-axis representing the reduced strain rate, which accounts for temperature differences.

The findings reveal that the interface normal strength is neither time- nor temperature-dependent. In contrast, the matrix strength demonstrates both time and temperature dependence.

2.3.2 Time and Temperature Dependence of Interface "Shear" Strength

In the previous section, it was found that the normal interface strength is not time and temperature dependent. In this part, the focus shifts to shear strength, which can be divided into two cases: combined stress of "shear and normal" and combined stress of "shear and compressive". This distinction is important, as pure shear stress rarely occurs in isolation. For the case of combined stress of "shear and normal", one article is introduced [9]. This study investigates the temperature dependence of interface strength under this combined stress state. The microbond test was conducted under varying temperature conditions, and the corresponding numerical simulation was performed to identify the precise stress that initiated the complete interface failure, similar to the approach in Section 2.2.1. The conclusion of this study is that the interface strength is not temperature dependent.

For the case of combined stress of "shear and compressive", another article is introduced [4]. In this research, a single-fiber composite is used to investigate the time-dependent behavior of interface failure. The strain distribution in the carbon fiber is experimentally obtained using micro-Raman spectroscopy, with the results shown in Figure 2.8. The fiber stress recovery behavior around the fiber-breaking point changes with time, and the interface failure criterion under shear and compressive combined stress state is examined. The experimental procedure is described in the authors' work [15].

In order to numerically simulate the phenomenon discussed in the previous sections, finite element analysis was conducted using a viscoelastic constitutive relationship for matrix resin, stress recovery due to interface friction, and interface failure due to cohesive zone modeling [4]. In the commercial software ABAQUS, a quasi-parabolic criterion, as shown in Figure 2.9(a), was used. The numerical results demonstrate a very good agreement with experimental results. The key point to note is that when the interface has compressive stress, the interface shear strength is enhanced. By introducing this algorithm, the experimental behavior can be precisely simulated. The crucial finding is that while the interface "pure" shear strength is not time dependent, it appears to be time dependent because the interface compressive strength varies with time, causing the shear strength to seem time dependent. This is an essential aspect that can be applied to other cases. For example, consider a single-fiber fragmentation test conducted under various temperatures to evaluate the

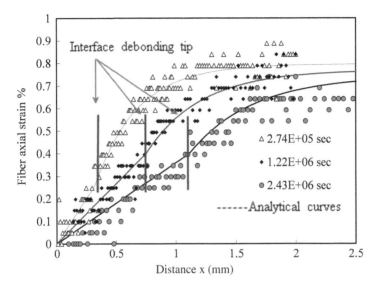

FIGURE 2.8 Strain distributions obtained from micro-Raman spectroscopy; the strain recovery profile regarding fiber-breaking point changes with time, that is, interface failure progresses with time, which is indicated by "interface debonding tip".

temperature dependence of interface strength. The results might show that interface strength decreases with temperature. However, this conclusion would not be accurate. If the temperature increases, the compressive stress at the interface decreases due to thermal deformation, causing the interface shear strength to decrease even though the interface pure shear strength remains unchanged. Accurately evaluating interface strength requires considering the failure envelope on the normal and shear stress plane. In conclusion, the evidence presented in this section supports the notion that interface strength is not time dependent, nor temperature dependent. Simultaneously, the "parabolic criterion" for the interface failure envelope in Figure 2.1 is suggested as an ideal interface failure criterion.

In addition, experimental data, as shown in Figure 2.10, suggest that shear strength is enhanced by compressive stress [6, 16–18], supporting the idea that interface strength may be time and temperature dependent. If the parabolic criterion is adopted, compressive stress is governed by the immediately neighboring matrix stress, which is inherently time and temperature dependent. Therefore, the interface shear strength may also change with changes in compressive stress, indicating that interface strength may indeed be time and temperature dependent. It is worth noting that research on the time and temperature

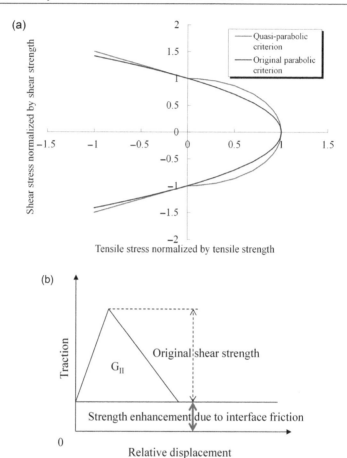

FIGURE 2.9 Characteristics of the cohesive element with the quasi-parabolic criterion. (a) Comparison of the failure criterion with the original parabolic criterion; (b) traction–separation behavior in shear with friction of the cohesive element.

dependence of interface strength typically focuses on interfacial shear strength, while interfacial normal strength may be less affected by such factors.

Several studies have suggested that the time and temperature dependence of interface strength is negligible, even in shear failure mode, using a quasi-parabolic criterion [4, 20]. This finding is important for understanding the durability of composite materials at the micromechanical level. Ultimately, a more comprehensive understanding of interface strength and its time and temperature dependence will be critical for accurate prediction of the failure behavior of composite materials.

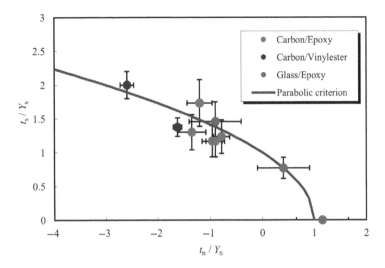

FIGURE 2.10   Stress ratio between normalized tensile and shear at interface failure for carbon fiber/epoxy [16, 19], carbon fiber/vinyl ester [4], and glass fiber/epoxy [16] with parabolic criterion as interface failure envelope.

2.4 APPROACH USING MOLECULAR DYNAMICS SIMULATION

Molecular simulations, such as molecular dynamics, have become increasingly efficient for evaluating interface mechanical properties [21–24], and they offer the possibility of designing interfaces more effectively than conducting numerous experimental investigations. In a study comparing experimental and numerical results in terms of mechanical properties of fiber/matrix interfaces, several specimens consisting of carbon fiber and various resins were prepared for microbond testing. Molecular-level interface modeling of graphene and the respective resins was performed, and interfacial energies were calculated on the basis of molecular dynamics. The orders of interfacial strengths obtained from experimental investigations and interfacial energies obtained from molecular simulations showed good agreement overall. The numerical results of density distribution near the graphene and resin molecular structures supported the relative order of the magnitude of the strength and interfacial energy values.

Epoxy resin, vinyl ester resin, and triA-X polyimide resin were used in the study, with carbon fiber represented by a graphene sheet (1,728 atoms per molecule). The molecular structures of these resins are shown in Figure 2.11. PolyParGen was used for resin modeling, and the OPLS-AA potential was employed. The molecular structures of the resin molecules were optimized using the B3LYP hybrid density functional with a 6–31G(d) basis set. The

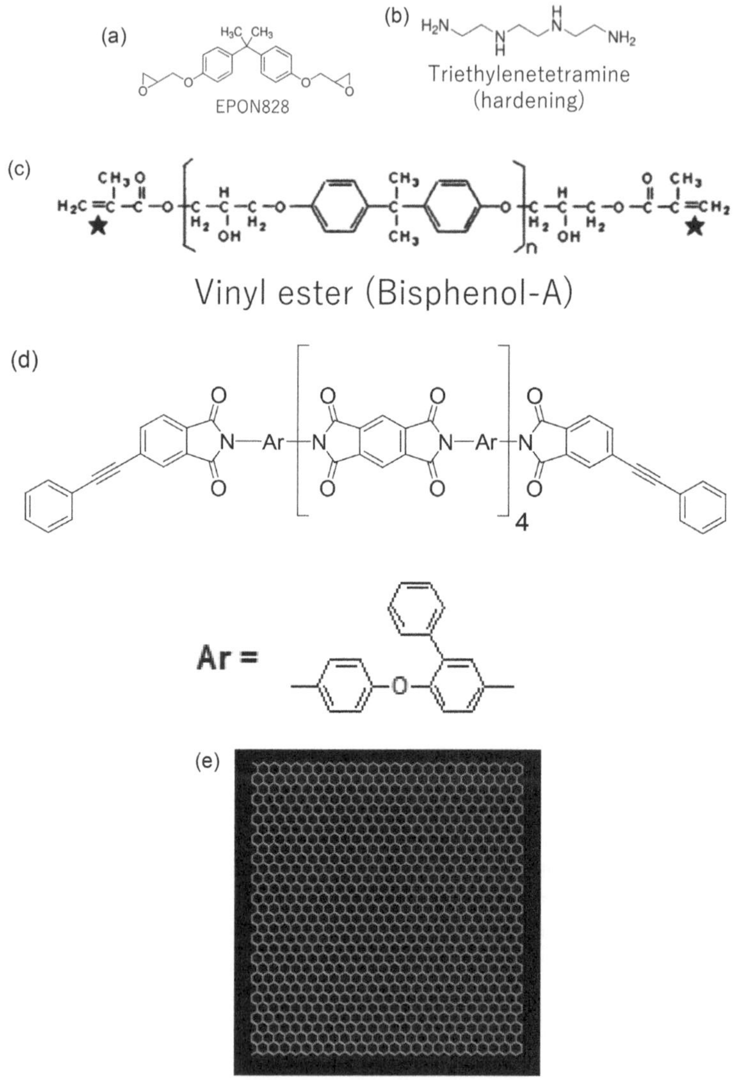

FIGURE 2.11 Molecular structure (a) epoxy molecule, (b) triethylenetetramine (curing agent), (c) vinyl ester molecule, (d) TriA-X polyimide, and (e) graphene sheet.

charge densities for each optimized structure were calculated at the PM3/6–31G(d) level. In this study, the oligomer state was assumed without considering crosslinking between resins. Although molecular crosslinking should be considered in reality, oligomers with low calculation costs were targeted as an initial trial. In the future, calculations should be performed considering crosslinking between resins.

As an interphase model, the resin molecules were randomly placed on cells with periodic boundary conditions above and below the three graphene sheets. The three sheets were stacked so that the resins would not attract each other across the graphene sheet, avoiding complications in calculating interfacial energy.

A periodic boundary condition was set for the cell in the x, y, and z directions. We used the general purpose software JOCTA (JSOL Corporation). For the interphase models, molecular dynamics calculations were performed using exabyte.io (MD calculation software). The time step used for the trajectory integration was 1 fs, and the procedure to obtain equilibrium configurations of the systems was as follows:

(1) After structural relaxation, MD simulation at 300 K was performed for 100 ps using the NVT ensemble.
(2) The system was compressed by raising the pressure in the Z direction to 10, 100, and 1000 MPa until the resin density reached approximately 1.0 g/cm^3.
(3) MD simulation at 300 K and 0.1 MPa was performed for 10 ns using the NPT ensemble.
(4) MD simulation at 300 K was performed for 10 ns using the NVT ensemble to reach equilibrium.

The Nose–Hoover thermostat and Parrinello–Rahman barostat were used.

We used the interfacial energy value per unit area (E_a) as a parameter to evaluate the binding strength of polymers to the graphene surface as defined in Eq. (2.3).

$$E_a = \left\{ E_{total} - \left(E_{re\sin} + E_{graphene} \right) \right\} / A \qquad \text{Eq. (2.3)}$$

where E_{total} is the total energy of the system, $E_{graphene}$ is the energy of the three graphene sheets, E_{resin} is the energy of polymer chains, and A is the graphene/polymer interfacial area. The calculated values of the interfacial energy are listed in Table 2.1. Negative interfacial energy indicates that the total energy of the system is lower than the sum of the energies of the isolated states, which is more stable and indicates that the interface is bonded.

TABLE 2.1 Comparisons Between Experiments With Results of Molecular Simulations [17]

POLYMER	INTERFACE ENERGY [J/M²]	INTERFACE STRENGTH OBTAINED BY EXPERIMENTS [MPa]
Epoxy	−0.15	74
Vinyl ester	−0.097	43
Polyimide (n = 4)	−0.19	130
Polyimide (n = 7)	−0.19	N/A

It should be noted that we could not directly compare the interfacial strength obtained experimentally and the interfacial energy obtained by molecular simulation. However, the strength values obtained experimentally followed the order: polyimide > epoxy > vinyl ester, and the calculated interfacial energy of the three systems followed the same order. If we assume that the interfacial strength and energy possess the same magnitude relation, we can conclude that this result is not inconsistent.

The calculated values of interfacial energy of the 7 monomer–oligomer TriA-X polyimide are listed in Table 2.1. As shown in Table 2.1, the 4 and 7 monomer–oligomer TriA-X polyimide exhibit similar interfacial energy values. Thus, although we did not perform microbonding tests on the 7 monomer-oligomer TriA-X polyimide, it is likely that the interfacial strength of the 7 monomer–oligomer polyimide specimen will be approximately 130 MPa.

Thus, this section introduces a possibility of interface strength evaluation by molecular dynamics simulation. Nowadays, the mechanism of interface mechanical properties is about to be clarified, described in this chapter.

REFERENCES

[1] J. Koyanagi, H. Hatta, M. Kotani, H. Kawada, A comprehensive model for determining tensile strengths of various unidirectional composites, Journal of Composite Materials 43(18) (2009) 1901–1914.

[2] J. Koyanagi, S. Ogihara, H. Nakatani, T. Okabe, S. Yoneyama, Mechanical properties of fiber/matrix interface in polymer matrix composites, Advanced Composite Materials 23 (2014) 551–570.

[3] J. Koyanagi, G. Kiyota, T. Kamiya, H. Kawada, Prediction of creep rupture in unidirectional composite: Creep rupture model with interfacial debonding and its propagation, Advanced Composite Materials: The Official Journal of the Japan Society of Composite Materials 13(3–4) (2004) 199–213.

[4] J. Koyanagi, A. Yoshimura, H. Kawada, Y. Aoki, A numerical simulation of time-dependent interface failure under shear and compressive loads in single-fiber composites, Applied Composite Materials 17(1) (2010) 31–41.

[5] J. Koyanagi, H. Nakatani, S. Ogihara, Comparison of glass-epoxy interface strengths examined by cruciform specimen and single-fiber pull-out tests under combined stress state, Composites Part A: Applied Science and Manufacturing 43(11) (2012) 1819–1827.

[6] S. Ogihara, J. Koyanagi, Investigation of combined stress state failure criterion for glass fiber/epoxy interface by the cruciform specimen method, Composites Science and Technology 70(1) (2010) 143–150.

[7] J. Koyanagi, S. Ogihara, Temperature dependence of glass fiber/epoxy interface normal strength examined by a cruciform specimen method, Composites Part B: Engineering 42(6) (2011) 1492–1496.

[8] M. Sato, E. Imai, J. Koyanagi, Y. Ishida, T. Ogasawara, Evaluation of the interfacial strength of carbon-fiber-reinforced temperature-resistant polymer composites by the micro-droplet test, Advanced Composite Materials 26(5) (2017) 465–476.

[9] M. Sato, J. Koyanagi, X. Lu, Y. Kubota, Y. Ishida, T.E. Tay, Temperature dependence of interfacial strength of carbon-fiber-reinforced temperature-resistant polymer composites, Composite Structures 202 (2018) 283–289.

[10] J. Koyanagi, P.D. Shah, S. Kimura, S.K. Ha, H. Kawada, Mixed-mode interfacial debonding simulation in single-fiber composite under a transverse load, Journal of Solid Mechanics and Materials Engineering 3(5) (2009) 796–806.

[11] Y. Miyano, M. Nakada, M.K. McMurray, R. Muki, Prediction of flexural fatigue strength of CRFP composites under arbitrary frequency, stress ratio and temperature, Journal of Composite Materials 31(6) (1997) 619–638.

[12] Y. Miyano, M. Nakada, Time and temperature dependent fatigue strengths for three directions of unidirectional CFRP, Experimental Mechanics 46(2) (2006) 155–162.

[13] J. Koyanagi, S. Yoneyama, K. Eri, P.D. Shah, Time dependency of carbon/epoxy interface strength, Composite Structures 92(1) (2010) 150–154.

[14] J. Koyanagi, S. Yoneyama, A. Nemoto, J.D.D. Melo, Time and temperature dependence of carbon/epoxy interface strength, Composites Science and Technology 70(9) (2010) 1395–1400.

[15] J. Koyanagi, H. Hatta, F. Ogawa, H. Kawada, Time-dependent reduction of tensile strength caused by interfacial degradation under constant strain duration in UD-CFRP, Journal of Composite Materials 41(25) (2007) 3007–3026.

[16] S. Kimura, J. Koyanagi, D. Yamamoto, H. Kawada, A novel method for evaluation of fiber strength using fragmentation test, Japanese Journal of Experimental Mechanics 6 (2006) 122–127 (In Japanese).

[17] J. Koyanagi, H. Hatta, F. Ogawa, H. Kawada, Time-dependent reduction of tensile strength caused by interfacial degradation under constant strain duration in UD-CFRP, Journal of Composite Materials 41(25) (2016) 3007–3026.

[18] J. Koyanagi, J. Kawal, S. Ogihara, K. Watanabe, Carbon fiber/matrix interfacial shear strength evaluated by single fiber. pull-out test considering effects of resin meniscus, Japanese Journal of Experimental Mechanics 2010 (2010) 407–412 (In Japanese).

[19] A. Straub, M. Slivka, P. Schwartz, A study of the effect of time and temperature on the fiber/matrix interface using the microbond test, Composites Science and Technology 57 (1997) 991–994.

[20] M. Sato, J. Koyanagi, X. Lu, Y. Kubota, Y. Ishida, T.E. Tay, Temperature dependence of interfacial strength of carbon-fiber-reinforced temperature-resistant polymer composites, Composite Structures 202 (2018) 283–289.

[21] T. Niuchi, J. Koyanagi, R. Inoue, Y. Kogo, Molecular dynamics study of the interfacial strength between carbon fiber and phenolic resin, Advanced Composite Materials 26(6) (2017) 569–581.

[22] J. Koyanagi, N. Itano, M. Yamamoto, K. Mori, Y. Ishida, T. Bazhirov, Evaluation of the mechanical properties of carbon fiber/polymer resin interfaces by molecular simulation, Advanced Composite Materials 28(6) (2019) 639–652.

[23] S. Naito, J. Koyanagi, T. Komukai, T. Uno, Analysis of three-phase structure of epoxy resin/CNT/Graphene by molecular simulation, Polymers 12(8) (2020).

[24] M. Morita, Y. Oya, N. Kato, K. Mori, J. Koyanagi, Effect of electrostatic interactions on the interfacial energy between thermoplastic polymers and graphene oxide: A molecular dynamics study, Polymers 14(13) (2022).

Constitutive Relationship of Resin Matrix Including Viscoelasticity and Damage

3

3.1 NONLINEAR VISCOELASTIC CONSTITUTIVE EQUATION CONSIDERING IRRECOVERABLE STRAIN FOR VINYL ESTER RESIN

3.1.1 Introduction

Polymer matrix composites (PMCs) are currently considered to be the next-generation structural material for use in the aerospace and construction industries, owing to their superior specific strength and stiffness compared to metals. However, unlike metals, PMCs exhibit time-dependent behaviors such as creep and stress relaxation, even at room temperature, and are known to fail prematurely relative to their design life. A number of previous studies [1–10] have investigated the creep behavior of composite materials, revealing that the viscoelastic deformation of the matrix strongly influences the

DOI: 10.1201/9781003371137-3

creep of composite materials. Nevertheless, these studies aimed to simplify the formulation of creep strain in composite materials by expressing the viscoelastic deformation of the matrix in a very simple formula, which may not accurately represent actual materials' viscoelastic behavior. Consequently, it is challenging to predict the creep deformation of composite materials with such formulations, particularly for long-term deformation. Thus, to assess the long-term reliability of composite materials, it is essential to clarify the viscoelastic deformation of the matrix, which is the dominant factor.

Polymer materials typically display viscoelastic behavior, wherein their strain and stress responses change over time depending on their loaded stress and strain history. At relatively low levels of stress or strain, the materials exhibit linear behavior. However, at higher levels of stress, they exhibit nonlinear behavior. Schapery et al. [11–13] have conducted research on the nonlinear viscoelastic constitutive equation of polymer materials by adding a nonlinear parameter to the conventional linear viscoelastic stress–strain constitutive equation. This enables the formulation of the nonlinear viscoelastic deformation under stress levels higher than a certain threshold. However, in actual material viscoelastic deformation, there is a permanent deformation that occurs in addition to the deformation that can recover to zero after the stress is removed. Schapery's formulation does not account for this permanent deformation, which does not fully satisfy the nonlinear viscoelastic deformation. Furthermore, neglecting the formulation of this permanent strain fails to accurately describe the viscoelastic deformation of the resin under long-term or high stress conditions. Therefore, it is necessary to formulate the permanent strain in order to properly discuss the long-term reliability of nonlinear viscoelastic materials.

In this study, we performed creep-recovery tests on vinyl ester resin and used the methods of Cardon et al. [14–16] to calculate the nonlinear parameters in Schapery's formulation. Furthermore, we investigated the permanent strain that occurs during the test, which has not been thoroughly studied thus far, in order to formulate the nonlinear viscoelastic deformation of the resin more accurately.

3.1.2 Theoretical Background

The constitutive equation for linear viscoelasticity, incorporating the superposition principle to account for stress history, is given by the Boltzmann superposition integral. This equation allows for the strain of a viscoelastic material to be described as follows:

$$\varepsilon(t) = \int_{0^-}^{t} D(t-\tau) \frac{d\sigma}{d\tau} d\tau \qquad\qquad \text{Eq. (3.1)}$$

However, in actual polymer materials, nonlinear viscoelastic deformation occurs when relatively large stresses are applied, where the creep compliance becomes dependent on stress. Schapery et al. [11–13] have thermodynamically derived the following single integral nonlinear viscoelastic constitutive equation:

$$\varepsilon(t) = g_0 D_0 \sigma + g_1 \int_0^t \Delta D(\psi - \psi') \frac{dg_2 \sigma}{d\tau} d\tau \qquad \text{Eq. (3.2)}$$

$$\psi \equiv \int_0^t dt' / a_\sigma, \ \psi' \equiv \int_0^\tau dt' / a_\sigma$$

This constitutive equation expresses the stress-dependent nonlinearity of creep deformation by using parameters that depend on stress.

Let us now discuss Schapery's constitutive equation (Eq. 3.2) used in this study. Here D_0 and $\Delta D(t)$ are the initial creep compliance and creep compliance increment up to time t, respectively, in linear viscoelasticity. The parameters a_σ, g_0, g_1, and g_2 are functions of stress and are called nonlinear parameters, which express the stress-dependent nonlinearity of creep deformation. The parameters ψ and ψ' represent the effects of stress change and the time at which they occur on the deformation and are called equivalent times. As these nonlinear parameters become constants (=1) at low stress, Eq. (3.2) reduces to the linear constitutive equation (Eq. 3.1). It has been proposed to use these parameters to describe the behavior of strain under various stress histories. In this study, we adopt Schapery's constitutive equation as a basis for representing nonlinear viscoelastic deformation, which is effective even in strongly nonlinear cases and can accommodate various stress histories.

Cardon et al. considered that in creep tests, when the stress is unloaded, the strain does not fully recover to zero, and there is residual strain that accumulates during the creep process. Based on this observation, Cardon proposed that the creep strain is the sum of the recoverable strain that can be returned to zero after stress unloading and the permanent strain that cannot be recovered. Thus, Cardon formulated the following equation by adding a permanent strain term, ε_{pd}, to Eq. (3.2),

$$\varepsilon(t) = g_0 D_0 \sigma + g_1 \int_0^t \Delta D(\psi - \psi') \frac{dg_2 \sigma}{d\tau} d\tau + \varepsilon_{pd}(t) \qquad \text{Eq. (3.3)}$$

$$\psi \equiv \int_0^t dt' / a_\sigma, \psi' \equiv \int_0^\tau dt' / a_\sigma$$

The increase in creep compliance in the linear viscoelastic regime is assumed to follow the following equation

$$\Delta D(t) = Ct^n \qquad \text{Eq. (3.4)}$$

Here C and n are creep constants and $0 < n < 0.5$. Cardon et al. [14–16] proposed a method to individually determine each parameter through creep-recovery tests. First, they assumed that there is a boundary stress, σ_c, between linear viscoelasticity and nonlinear viscoelasticity, below which all nonlinear parameters are equal to 1 and above which each nonlinear parameter varies depending on the stress. Figure 3.1 illustrates a schematic diagram of the strain behavior in creep-recovery tests.

When time = 0, a stress σ_0 is applied, and then it is removed at t_a. The stress history is

$$\sigma(t) = \begin{cases} \sigma_0 & 0 < t < t_a \\ 0 & t > t_a \end{cases}$$

Eq. (3.5)

According to Eqs (3.3), (3.4), and (3.5), strain during creep loading, $\varepsilon_c(t)$, and recovering strain after unloaded, $\varepsilon_r(t)$, are expressed as follows, respectively.

$$\varepsilon_c(t) = g_0 D_0 \sigma_0 + g_1 g_2 C \left(\frac{t}{a_\sigma} \right)^n \sigma_0 + \varepsilon_{pd}(t)$$

Eq. (3.6)

$0 < t < t_a$

$$\varepsilon_r(t) = \frac{\Delta\varepsilon_c - \varepsilon_{pd}(t_a)}{g_1} \left[(1 + a_\sigma \lambda)^n - (a_\sigma \lambda)^n \right] + \varepsilon_{pd}(t_a)$$

$t > t_a$

Eq. (3.7)

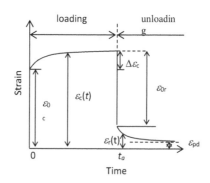

FIGURE 3.1 Typical strain vs. time curve for creep-recovery test.

Here, $\Delta\varepsilon_c$ includes irrecoverable deformation until t_a, and λ is dimensionless time as expressed by

$$\Delta\varepsilon_c = g_1 g_2 C \left(\frac{t_a}{a_\sigma}\right)^n \sigma_0 + \varepsilon_{pd}(t_a) \qquad \text{Eq. (3.8)}$$

$$\lambda \equiv \left.(t - t_a)\middle/ t_a\right.$$

In the range of linear viscoelastic, $\varepsilon_c(t)$ and $\varepsilon_r(t)$ are as follows, respectively.

$$\varepsilon_c(t) = D_0 \sigma_0 + C t^n \sigma_0 + \varepsilon_{pd}(t) \qquad \text{Eq. (3.9)}$$

$$0 < t < t_a$$

$$\varepsilon_r(t) = (\Delta\varepsilon_c - \varepsilon_{pd})\left[(1+\lambda)^n - (\lambda)^n\right] + \varepsilon_{pd}(t_a) \qquad \text{Eq. (3.10)}$$

$$\Delta\varepsilon_c = C t_a{}^n \sigma_0 + \varepsilon_{pd}(t_a) \, t > t_a$$

There is a relationship between the initial elastic strain ε_{0c}, increment of creep strain $\Delta\varepsilon_c$, and recovery strain instantaneously respond at unloading ε_{0r} as follows.

$$\Delta\varepsilon_0 = \varepsilon_{0r} - \varepsilon_{0c} \qquad \text{Eq. (3.11)}$$

Next, ε_{0r} is difference between $\varepsilon_c(t)$ and $\varepsilon_r(t)$ at t_a so that $\Delta\varepsilon_0$ can be formulated as following [14].

$$\Delta\varepsilon_0 = g_2(g_1 - 1)C\left(\frac{t_a}{a_\sigma}\right)^n \sigma_0 \qquad \text{Eq. (3.12)}$$

From Eqs (3.12) and (3.8), g_1 can be determined as

$$g_1 = \frac{\Delta\varepsilon_c - \varepsilon_{pd}(t_a)}{\Delta\varepsilon_c - \varepsilon_{pd}(t_a) - \Delta\varepsilon_0} \qquad \text{Eq. (3.13)}$$

Further, a_σ is determined by a comparison of Eq. (3.7) with experimental curves. Next, let's consider g_2. Cardon et al. [15] proposed a method for determining g_2 as a function of stress by using experimental results where the loading stress is linear, that is, below the boundary stress σ_c between linear and nonlinear viscoelasticity. Creep-recovery tests were performed for both linear and nonlinear stresses, and g_2 for stresses above σ_c was formulated using the obtained nonlinear $\Delta\varepsilon_0$, linear creep strain increment $\Delta\varepsilon_c$, and permanent strain ε_{pd}.

$$g_2 = \frac{\Delta\varepsilon_{0(nl)}}{\Delta\varepsilon_{c(l)} - \varepsilon_{pd(l)}(t_a)} \frac{a_\sigma^n}{(g_1-1)} \frac{\sigma_{0(l)}}{\sigma_{0(nl)}} \qquad \text{Eq. (3.14)}$$

In Equation (3.14), subscripts (l) and (nl) are added to indicate linear and nonlinear strain and stress results, respectively, to distinguish between linear and nonlinear test results. Equation (3.14) determines g_2 as a function of stress for both linear and nonlinear creep-recovery tests. However, permanent strain is calculated by fitting Eqs (3.7) and (3.10) to the creep-recovery curve. The parameter g_0, which pertains to nonlinear elasticity, is determined by the change in initial elastic strain under stress loading. However, in these studies, no examination or formulation has been performed regarding permanent strain, and thus the formula is incomplete with regard to predicting long-term or high-stress strain. Therefore, in this study, we conducted various creep-recovery tests at different stresses and times and investigated the stress–time dependence of permanent strain, aiming to provide a more complete understanding of the phenomenon.

3.2 CREEP-RECOVERY TESTS

Table 3.1 shows the resin blending ratios, and Figure 3.2 shows the shapes of the test pieces. The test pieces were fabricated by cutting the resin after it was

TABLE 3.1 Properties of Vinyl Ester Resin

	CHEMICALS	COMPOUNDING RATIO
Active component	Ripoxy R802	100
Curing agent	Methyl ethyl ketone peroxide	0.9
Accelerator	Cobalt naphthenate	0.3

cured to the desired shape. The curing conditions for the vinyl ester resin were three hours at room temperature followed by two hours at 80°C. However, the glass transition temperature of the vinyl ester resin is around 95°C.

The creep-recovery test was conducted using a six-point creep testing machine at a constant temperature of 50°C. Prior to beginning the test, the test pieces were left in the test environment for several hours. The stress was then loaded, and the stress σ_0 was removed at time ta. The loading stress σ0 ranged from 5 to 45 MPa, and the creep time t_a was 0.6, 1.0, 2.0, 6.0, and 24.0 hours, while the recovery time was three times the creep time. Strain measurement was carried out using a displacement meter, with a flat section of 50 mm in length in Figure 3.2 used as the distance between reference points.

Figure 3.3 shows the changes in creep compliance at each stress obtained from creep-recovery tests with creep time of t_a 6.0 and 24.0 hours, respectively. However, the tests could not be performed at stress levels above 40 MPa because the specimens ruptured in a short period of time. Moreover, in tests with stress levels below 10 MPa, despite being loaded with stress, many tests exhibited unstable behavior such as specimen contraction over time. This is believed to be due to the fact that the viscoelastic behavior of polymer materials is highly sensitive to environmental factors such as humidity and temperature, and this effect is particularly pronounced in the strain response at low stress levels.

Based on the results of the creep-recovery tests at creep times of t_a = 6.0 and 24.0 hours shown in Figure 3.3, it is evident that vinyl ester resin exhibits nonlinear viscoelastic deformation with creep compliance depending on stress. It is observed from Figure 3.3 that the most significant change in creep compliance occurs between 26 MPa and 28 MPa, and nonlinear viscoelasticity is particularly prominent at stresses above 28 MPa. In Figure 3.3(a), a slight stress dependence of creep compliance is observed at stresses below 26 MPa. However, such stress dependence is hardly observed in Figure 3.3(b), and at 26 MPa, the creep compliance is lower than at lower stresses, indicating the presence of experimental errors in the obtained creep compliance values. This trend is observed in many other test results, and the viscoelastic behavior of

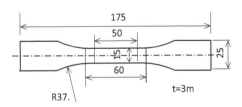

FIGURE 3.2 Specimen geometry.

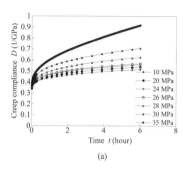
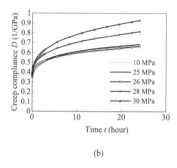

(a)

(b)

FIGURE 3.3 Experimental results for creep compliance for (a) six hours and (b) 24 hours.

polymer materials is very sensitive to environmental factors such as humidity and individual differences of the test specimens, inevitably resulting in errors in the obtained data. Therefore, it is concluded that the deviation of creep compliance at stresses below 26 MPa in Figure 3.3 is due to experimental errors or individual differences of the test specimens. On the basis of these results, we consider stresses above 28 MPa to exhibit significant stress dependence of creep compliance, indicating nonlinear viscoelasticity, and determine the linear/nonlinear boundary stress σ_c to be 26 MPa in this study.

They also determined the creep exponent n to be 0.27 by fitting Eq. (3.10) to the experimental results at stresses below 26MPa and determined the nonlinear parameters a_σ, g_1, and g_2, at each stress from the creep-recovery test results using Equations (3.7), (3.13), and (3.14), respectively, assuming a value of 1 for a_σ and g_2 in the linear range. The authors plotted the values and theoretical lines of the nonlinear parameters at each stress, as well as their error ranges reflecting experimental error and individual differences in creep compliance in Figures 3.5 and 3.6.

The nonlinear parameters are formulated using the following equations [16].

$$G_i = \begin{cases} 1 & \text{for } \sigma < \sigma_c \\ \dfrac{1-k_i}{1+\dfrac{x-\mu}{1-x}\exp\left(\dfrac{x-\mu}{1-x}\right)}+k_i & \text{for } \sigma > \sigma_c \end{cases} \qquad \text{Eq. (3.15)}$$

Here, G_i represents nonlinear parameters a_σ, g_1, g_2, for example, a_σ is expressed as G_σ. Also, $x = \sigma/\sigma_u$ ανδ $\mu = \sigma_c/\sigma_u$, where σ_u is ultimate strength under extremely rapid test and is assumed to be 65 MPa in this study. Further,

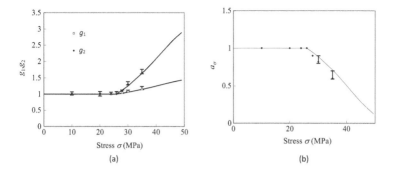

FIGURE 3.4 Nonlinear factors (a) g_1, g_2, and (b) a_σ.

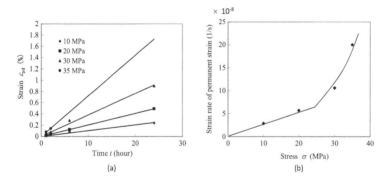

FIGURE 3.5 Relationship between (a) permanent strain and time at each stress and (b) permanent strain rate and stress.

asymptote values of each parameter at the ultimate strength k_i is 1.6, 3.2, 0, for g_1, g_2, a_σ, respectively. From Figure 3.4, it is clear that the nonlinear parameters vary monotonically with the applied stress and depend on the stress load exceeding 26 MPa. For the other nonlinear parameter, g_0, the variation with stress load, was found to be small for initial creep compliance, and thus it was assumed to be 1 in this study. The initial creep compliance in the linear range was determined to be 0.33 [1/GPa]. As a result, the nonlinear parameters a_σ, g_1, and g_2 in Schapery's constitutive equation were determined as functions of stress load.

Next, we will examine the permanent strain. Figure 3.5(a) shows the relationship between permanent strain ε_{pd} and time for each stress in the creep-recovery test. It can be seen from Figure 3.5(a) that permanent strain increases with time and stress. Assuming a linear relationship between ε_{pd} and time and a nonlinear relationship between ε_{pd} and stress, and assuming that the permanent strain can be described as a simple dashpot, the following relationship between stress and strain holds.

$$\sigma = \eta(\sigma)\frac{d\varepsilon_{pd}}{dt} \qquad\qquad \text{Eq. (3.16)}$$

Here η is viscosity and σ is applied stress. Figure 3.5(a) shows the relationship between permanent strain and time, and Figure 3.5(b) shows the relationship between strain rate of permanent one and applied stress.

Next, we calculate the flow rate ϕ from the strain rate and define a new nonlinear parameter g_3 that depends on stress for permanent strain, similar to recoverable strain.

$$\phi = \phi_0 g_3, \phi = 1/\eta \qquad\qquad \text{Eq. (3.17)}$$

The value of ϕ_0, the flow rate at 10 MPa, was found to be $2.82 \times 10-9$ [1/(MPa·s)]. Note that ϕ_0 is the reciprocal of the viscosity coefficient η_0 at 10 MPa. Figure 3.6 shows the experimental results and theoretical curve. The theoretical curve was calculated using the formula with the asymptotic value of $k_i = 5$ at the ultimate strength. It can be seen from Figure 3.6 that the newly defined nonlinear parameter has a linear/nonlinear boundary stress at 26 MPa, similar to Schapery's constitutive equation, and that permanent strain can be

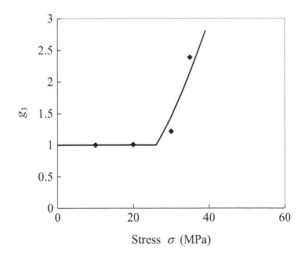

FIGURE 3.6   Nonlinear factor g_3 as a function of stress.

described by such a simple nonlinear dashpot. Therefore, a constitutive equation for permanent deformation was formulated, taking into account the time and stress dependence of permanent strain. The flow rate at a stress of 10 MPa is $2.82 \times 10-9$ 1/MPa·s. Here, ϕ_0 is the reciprocal of the viscosity coefficient η_0 at 10 MPa.

Figure 3.6 shows the experimentally determined value of g_3 and the theoretical curve, where the theoretical curve is obtained using Eq. (3.15) with the asymptotic value of $k_i = 5$ at the ultimate strength. From Figure 3.6, it is found that the newly defined nonlinear parameter has a linear/nonlinear boundary stress at 26 MPa, similar to the nonlinear parameter of Schapery's constitutive equation, and that the permanent strain can be described by such a simple nonlinear dashpot. Therefore, the constitutive equation for permanent deformation, which takes into account the time and stress dependence of the permanent strain, has been formulated.

Finally, by fitting Eq. (3.10) to the creep-recovery test results, the creep constant C was determined to be $1.20 \times 10-5$. Therefore, the nonlinear viscoelastic constitutive Eq. (3.3) taking into account the permanent strain was formulated by following Eq. (3.18).

$$\varepsilon(t) = g_0 D_0 \sigma + g_1 \int_0^t \Delta D(\psi - \psi') \frac{dg_2 \sigma}{d\tau} d\tau + \frac{\sigma \cdot t}{\eta_0} \cdot g_3 \qquad \text{Eq. (3.18)}$$

Figure 3.7 compares the theoretical curves obtained from the nonlinear viscoelastic constitutive equation formulated in this study with the actual strain results obtained from (a) creep recovery test, (b) multiple step creep recovery test, and (c) constant strain rate tests The figures show that the proposed constitutive equation accurately represents the viscoelastic deformation of the resin in both the linear and nonlinear stress regimes.

3.3 ENTROPY-BASED FAILURE CRITERION FOR POLYMER MATERIALS

The prior discourse concerns the analytical framework for the nonlinear viscoelastic properties of polymeric substances. This segment emphasizes the time-dependent failure criteria for addressing the prolonged durability of CFRPs. A time-dependent failure formulation is crucial for this objective. In this context, the entropy-damage criterion is particularly noteworthy due to its extensive applicability to viscoelastic media. This criterion, based on

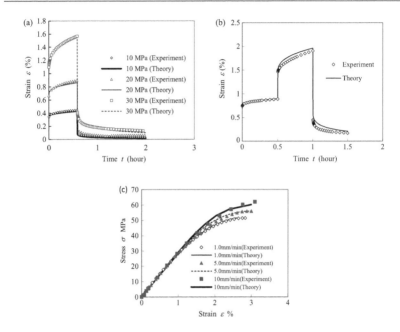

FIGURE 3.7 Comparisons between theoretical curves and experimental results for (a) creep recovery test, (b) multiple step creep recovery test, and (c) constant strain rate tests.

thermodynamic principles, is considered a vital instrument for examining the endurance of polymeric materials. Nonetheless, forecasting localized failure in intricate-shaped structural CFRPs through analytical solutions can be quite arduous. Therefore, a technique that integrates both nonlinear viscoelastic constitutive equations and the entropy failure criterion into computational simulations is indispensable. The authors introduce a computational model [17] that provides an accessible means to accomplish this aim. In this segment, we elaborate on such computational models comprehensively. Concerning the deformation of a continuous medium, internal variables like strain and temperature change depending on location and time, making the thermodynamic state of the medium nonequilibrium and heterogeneous. Consequently, conventional thermodynamics, grounded in the assumptions of homogeneity and equilibrium states, is not applicable for describing these thermodynamic processes. To tackle this challenge, the local states principle is postulated in nonuniform and nonequilibrium thermodynamics. According to this principle, the thermodynamic state of any point in the medium at any given time is entirely defined by the instantaneous state variables at that point, which are connected by the same equations as for a homogeneous and equilibrium system. This principle suggests that the thermodynamic assumptions of a moving continuous medium advance as a continuum of equilibrium states. Thus, within the scope of irreversible thermodynamic processes, the internal variables at each point as a function of time facilitate the treatment of the medium's dynamics, even though the system's current state relies on its historical background. Numerous researchers have employed this entropy damage criterion [18–29]. In this segment, we present an introduction to the utilization of the entropy damage criterion in computational simulations.

The entropy is defined as follows.

$$\gamma^{\mathrm{cr}} = \int_0^{t^{\mathrm{f}}} \frac{Q}{T} \, dt \qquad\qquad \text{Eq. (3.19)}$$

Here t^{f} is time at failure, T is absolute temperature, Q is dissipated energy consisting of work consumed for inelastic deformation and material damage, and U^{e} is elastic strain energy. For example, if we assume a usual stress–strain curve shown in Figure 3.8, obtained from tensile test of coupon resin specimen, Q and U are defined by these areas as shown in this figure. Figure 3.9 shows schematic overview for a polymer failure and void, which is equivalent to damage, nucleation, and growth. As the void grows, the entropy value is also elevated. When the entropy value reaches some specific one, the material fails. This is entropy failure criterion.

The viscoelastic behavior is expressed as the numerical model as shown in Figure 3.10.

The total strain can be divided into viscoelastic potion and viscoplastic portion. In the present study, the viscoplastic strain is modeled by a nonlinear dashpot, as shown in Figure 3.1. Therefore, the viscoplastic strain can be expressed through the following simple equation.

$$\varepsilon_{ij}^{\mathrm{vp}} = \int_0^t H_{ijkl}^{\mathrm{vp}}{}^{-1} \sigma_{kl} \, dt \qquad\qquad \text{Eq. (3.20)}$$

Here H can be written in a matrix form as follows:

$$\boldsymbol{H}^{\mathrm{vp}} = \frac{\eta_{\mathrm{vp}}}{(1+v)(1-2v)} \begin{bmatrix} (1-v) & v & v & 0 & 0 & 0 \\ v & (1-v) & v & 0 & 0 & 0 \\ v & v & (1-v) & 0 & 0 & 0 \\ 0 & 0 & 0 & \dfrac{1-2v}{2} & 0 & 0 \\ 0 & 0 & 0 & 0 & \dfrac{1-2v}{2} & 0 \\ 0 & 0 & 0 & 0 & 0 & \dfrac{1-2v}{2} \end{bmatrix} \qquad \text{Eq. (3.21)}$$

Here ν is Poisson's ratio, and η_{vp} is determined by the following equation:

$$\eta_{vp} = \frac{\eta_0 \left(1 + e^{\beta \left(\frac{\varepsilon_{eqv}^{vp}}{\sigma_{eqv}} \right)^n} \right)}{\left(1 + e^{\alpha \left(\sigma_{eqv} - \sigma_0 \right)} \right)}$$

Eq. (3.22)

Here subscript "eqv" is the equivalent value of the current state, η_0 is the initial value of α, β, and σ_0 and n are specific constants determined on the basis of a comparison between the experimental and analytical values. As shown in Eq. (3.22), η_{vp} is expressed as a function of the current equivalent viscoplastic strain and equivalent stress. The value of the denominator increases nonlinearly with the current stress when it exceeds a specific stress value of σ_0. The value of the numerator increases nonlinearly with the current viscoplastic strain and decreases with the current stress.

Viscoelastic strain can be obtained using stress tensor.

$$\varepsilon_{ij}^{ve} = \varepsilon_{ij}^{total} - \int_0^t \left(H_{ijkl}^{irr} \right)^{-1} \sigma_{kl} dt$$

Eq. (3.23)

Also, the stress is determined by convolution integral based on the relaxation modulus $E(t)$ and viscoelastic strain history with damage parameter D.

$$\sigma_{ij}(t) = (1-D) \int_0^t E_{ijkl}(t-t') \frac{g d\varepsilon_{kl}^{ve}}{dt'} dt'$$

Eq. (3.24)

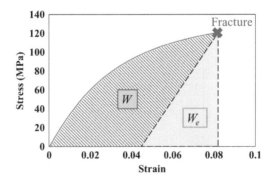

FIGURE 3.8 Definition of Q and U for the case of usual stress–strain curve obtained from usual tensile test of resin coupon specimen.

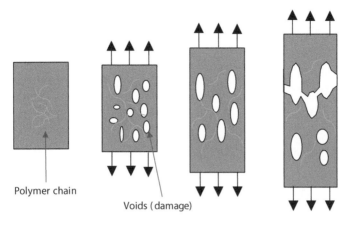

FIGURE 3.9 Schematic overview for damage nucleation and growth with tensile loading.

Here, the relaxation modulus $E(t)$ is defined as

$$E_{ijkl}(t) = \sum_{n=1}^{15} E_{ijkl}^n e^{-tE^n / \eta^n}$$

Eq. (3.25)

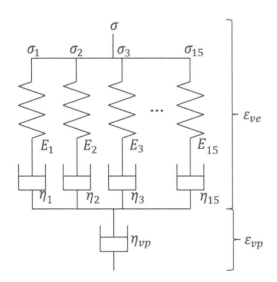

FIGURE 3.10 Viscoelastic–viscoplastic model consisting of 15 springs and 15 dampers.

Here superscript n means the number of Maxwell model and E and η are elasticity of spring and viscosity of dashpot, respectively. And g in Eq. (3.24) is also nonlinear parameter and defined as

$$g = 1 - \frac{1}{1 + e^{-\alpha\left(\frac{\sigma^{Mises} - \sigma_0}{\sigma_0}\right)}}$$

Eq. (3.26)

Here we suggest to define the material damage D which is determined by

$$D = \frac{D^{cr}}{\gamma^{cr}}\gamma$$

Eq. (3.27)

Here superscript "cr" means critical value. γ is entropy generation and defined by the value of dissipated energy divided by absolute temperature.

3.4 VALIDITY OF ENTROPY FAILURE CRITERION FOR RESIN FAILURE

3.4.1 Experimental Work

Sato et al. [28] examined the validity of entropy failure criterion for polyimide resin. The tensile failure tests are implemented by various strain-rates and various temperatures. The results are shown in Figure 3.11. Thus, a reasonable agreement can be obtained by the actual tests.

Stress is second-order tensor, and there are six components so that it is not simple to define the fracture of material. But entropy is scalar. In this study, we assume that the material fails when entropy reaches a critical value. It is desired to justify that the fracture entropy is the same independent of deformation mode, such as uniaxial, biaxial, triaxial, and shear stress state.

3.4.2 Numerical Work

Takase et al. (2021) [29] examine the fracture entropy with varying stress states by MD simulations, and the result is that the fracture entropy values are close values independent of the stress state. Full-atom simulation is implemented by GROMACS with 570,000 atoms. The assumed material is PA6. The fracture

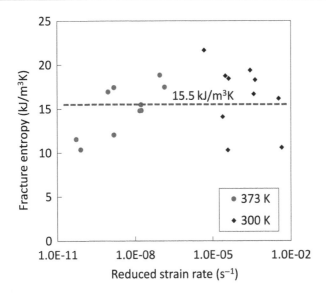

FIGURE 3.11 Entropy generation regarding tensile failure tests implemented at various strain-rates and various temperatures.

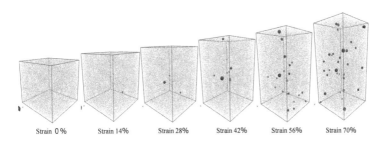

FIGURE 3.12 Void nucleation and growth with deformation under uniaxial tensile which is equivalent to triaxial tensile.

modes are uniaxial, biaxial, triaxial, and shear stress states. The entropy values are determined by dissipated energy divided by temperature, that is, mechanically measured entropy. In these simulations, the failure of covalent bonding is not considered, but this is reasonable for this material because the governing factor of the failure is entanglements of molecules. Furthermore, in this study, void size is proved as shown in Figure 3.12. For this result, the void size increases with the deformation, and entropy measured by thermodynamic point of view also increases simultaneously.

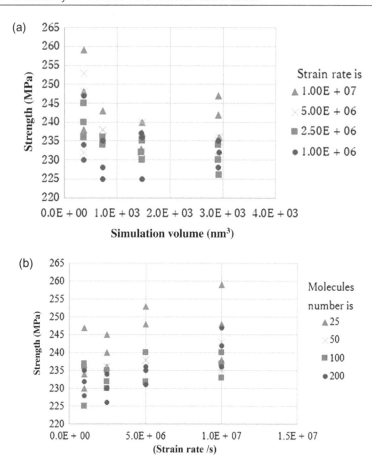

FIGURE 3.13 Relationship between failure strength and (a) simulation volume, the number of molecules and (b) strain rate.

3.5 STRENGTH OF RESIN PREDICTED BY MD SIMULATION

So far, it has been very difficult to numerically simulate the quantitative failure of polymer materials with molecular dynamics simulations. In most cases, the estimated strength is much greater than that in reality [30, 31]. This is partly because the failure of covalent bonding is usually not considered in molecular dynamics simulations. Koyanagi et al. (2020) [32] suggest a method for estimating the experimental strength of polymer materials. We are aware that there are relationships between the strength obtained by MD and the volume and strain rate, which mimic real phenomena. The strength of polymer materials decreases with their volume, based on Weibull statistics, and increases with an increase in strain rate, based on viscoelastic characteristics. Hence, we simulate the failure behavior in MD with various volumes and strain rates, as shown in Figure 3.13. Based on these results, the experimental strength is estimated by extrapolating the strength with the assumption of real volume and real strain rate. The estimated result is close to that obtained from real experiments. However, this method might only be applicable to thermoplastic resins in which the failure of covalent bonding does not play an important role in discussing the failure. For this case, we are now developing a new algorithm which simulates the failure of covalent bonding.

REFERENCES

[1] M. McLean, Creep deformation of metal-matrix composites, Composites Science and Technology 23(1) (1985) 37–52.

[2] W.A. Curtin, Theory of mechanical properties of ceramic-matrix composites, Journal of the American Ceramic Society 74(11) (1991) 2837–2845.

[3] P. Sofronis, R.M. McMeeking, The effect of interface diffusion and slip on the creep resistance of particulate composite materials, Mechanics of Materials 18(1) (1994) 55–68.

[4] Z.Z. Du, R.M. McMeeking, Creep models for metal matrix composites with long brittle fibers, Journal of the Mechanics and Physics of Solids 43(5) (1995) 701–726.

[5] N. Ohno, H. Kawabe, T. Miyake, M. Mizuno, A model for shear stress relaxation around fiber break in unidirectional composites and creep rupture analysis, Zairyo/Journal of the Society of Materials Science, Japan 47(2) (1998) 184–191.

[6] J. Koyanagi, G. Kiyota, T. Kamiya, H. Kawada, Prediction of creep rupture in unidirectional composite: Creep rupture model with interfacial debonding and its propagation, Advanced Composite Materials: The Official Journal of the Japan Society of Composite Materials 13(3–4) (2004) 199–213.

[7] Y. Miyano, M. Nakada, M.K. McMurray, R. Muki, Prediction of flexural fatigue strength of CRFP composites under arbitrary frequency, stress ratio and temperature, Journal of Composite Materials 31(6) (1997) 619–638.

[8] Y. Miyano, M. Nakada, Time and temperature dependent fatigue strengths for three directions of unidirectional CFRP, Experimental Mechanics 46(2) (2006) 155–162.

[9] J. Koyanagi, M. Nakada, Y. Miyano, Prediction of long-term durability of uni-directional CFRP, Journal of Reinforced Plastics and Composites 30(15) (2011) 1305–1313.

[10] J. Koyanagi, M. Nakada, Y. Miyano, Tensile strength at elevated temperature and its applicability as an accelerated testing methodology for unidirectional composites, Mechanics of Time-Dependent Materials 16(1) (2012) 19–30.

[11] R.A. Schapery, On the characterization of nonlinear viscoelastic materials, Polymer Engineering & Science 9(4) (1969) 295–310.

[12] R.A. Schapery, Nonlinear viscoelastic and viscoplastic constitutive equations based on thermodynamics, Mechanics Time-Dependent Materials 1(2) (1997) 209–240.

[13] Y.C. Lou, R.A. Schapery, Viscoelastic characterization of a nonlinear fiber-rein-forced plastic, Journal of Composite Materials 5(2) (1971) 208–234.

[14] S.P. Zaoutsos, G.C. Papanicolaou, A.H. Cardon, On the non-linear viscoelastic behaviour of polymer-matrix composites, Composites Science and Technology 58(6) (1998) 883–889.

[15] G.C. Papanicolaou, S.P. Zaoutsos, A.H. Cardon, Prediction of the non-linear vis-coelastic response of unidirectional fiber composites, Composites Science and Technology 59(9) (1999) 1311–1319.

[16] G.C. Papanicolaou, S.P. Zaoutsos, A.H. Cardon, Further development of a data reduction method for the nonlinear viscoelastic characterization of FRPs, Composites Part A: Applied Science and Manufacturing 30(7) (1999) 839–848.

[17] J. Koyanagi, K. Hasegawa, A. Ohtani, T. Sakai, K. Sakaue, Formulation of non-linear viscoelastic–viscoplastic constitutive equation for polyamide 6 resin, Heliyon 7(2) (2021).

[18] M. Naderi, M. Amiri, M.M. Khonsari, On the thermodynamic entropy of fatigue fracture, Proceedings of the Royal Society A: Mathematical, Physical and Engineering Sciences 466(2114) (2010) 423–438.

[19] H. Deng, A. Mochizuki, M. Fikry, S. Abe, S. Ogihara, J. Koyanagi, Numerical and experimental studies for fatigue damage accumulation of CFRP cross-ply laminates based on entropy failure criterion, Materials 16(1) (2023).

[20] J. Huang, H. Yang, W. Liu, K. Zhang, A. Huang, Confidence level and reliabil-ity analysis of the fatigue life of CFRP laminates predicted based on fracture fatigue entropy, International Journal of Fatigue 156 (2022).

[21] A. Imanian, M. Modarres, A thermodynamic entropy-based damage assess-ment with applications to prognostics and health management, Structural Health Monitoring 17(2) (2018) 240–254.

[22] A.P. Jirandehi, M.M. Khonsari, Microstructure-sensitive estimation of fatigue life using cyclic thermodynamic entropy as an index for metals, Theoretical and Applied Fracture Mechanics 112 (2021).

[23] H. Kagawa, Y. Umezu, K. Sakaue, J. Koyanagi, Numerical simulation for the tensile failure of randomly oriented short fiber reinforced plastics based on a vis-coelastic entropy damage criterion, Composites Part C: Open Access 10 (2023).

[24] J. Koyanagi, A. Mochizuki, R. Higuchi, V.B.C. Tan, T.E. Tay, Finite element model for simulating entropy-based strength-degradation of carbon-fiber-rein-forced plastics subjected to cyclic loadings, International Journal of Fatigue 165 (2022).

[25] M. Liakat, M.M. Khonsari, On the anelasticity and fatigue fracture entropy in high-cycle metal fatigue, Materials and Design 82 (2015) 18–27.

[26] B. Mohammadi, A. Mahmoudi, Developing a new model to predict the fatigue life of cross-ply laminates using coupled CDM-entropy generation approach, Theoretical and Applied Fracture Mechanics 95 (2018) 18–27.

[27] B. Mohammadi, M.M. Shokrieh, M. Jamali, A. Mahmoudi, B. Fazlali, Damage-entropy model for fatigue life evaluation of off-axis unidirectional composites, Composite Structures 270 (2021).

[28] M. Sato, K. Hasegawa, J. Koyanagi, R. Higuchi, Y. Ishida, Residual strength prediction for unidirectional CFRP using a nonlinear viscoelastic constitutive equation considering entropy damage, Composites Part A: Applied Science and Manufacturing 141 (2021).

[29] N. Takase, J. Koyanagi, K. Mori, T. Sakai, Molecular dynamics simulation for evaluating fracture entropy of a polymer material under various combined stress states, Materials 14(8) (2021).

[30] J. Koyanagi, N. Itano, M. Yamamoto, K. Mori, Y. Ishida, T. Bazhirov, Evaluation of the mechanical properties of carbon fiber/polymer resin interfaces by molecular simulation, Advanced Composite Materials 28(6) (2019) 639–652.

[31] T. Niuchi, J. Koyanagi, R. Inoue, Y. Kogo, Molecular dynamics study of the interfacial strength between carbon fiber and phenolic resin, Advanced Composite Materials 26(6) (2017) 569–581.

[32] J. Koyanagi, N. Takase, K. Mori, T. Sakai, Molecular dynamics simulation for the quantitative prediction of experimental tensile strength of a polymer material, Composites Part C: Open Access 2 (2020).

Composite Strength Estimation and Prediction Based on Micromechanics

4

4.1 FIBER-AXIAL TENSILE STRENGTH [1]

Koyanagi et al. [1] introduced a holistic model in 2009 for predicting the strengths of various composites. The study addresses the strength characterization of unidirectional composites, which is crucial in structural design. While numerous models exist for estimating unidirectional composite strength, their applicability remains restricted. The study introduces the Simultaneous Fiber-Failure (SFF) model, which calculates the diverse unidirectional composite strengths based on fiber strength, interfacial strength, and matrix strength. The SFF model examines whether adjacent fibers fail when a weak fiber fails and determines the number of simultaneous fiber failures, considering the interplay among fiber strength, interfacial strength, and matrix strength.

For a range of composites, such as glass and carbon-fiber-reinforced polymeric composite (GFRP, CFRP), ceramic matrix composite (CMC), and carbon/carbon composite (C/C), a unique empirical relationship between the number of simultaneous fiber failures and a coefficient representing the

DOI: 10.1201/9781003371137-4

interaction of component strengths can be derived. This number is formulated as a function of the coefficient and incorporated into the conventional Global Load Sharing (GLS) model. Ultimately, the unidirectional composite strength is expressed as a function of each composite component's strength. The SFF model delivers accurate predictions of various composite strengths, proving beneficial for optimal composite design, long-term assurance of composite materials, and the development of new composite systems.

The study also underscores the significance of matrix failure in determining unidirectional composite strength and acknowledges that the contribution of matrix failure to composite strength is still being debated. In conclusion, the SFF model offers a comprehensive approach to calculating the various unidirectional composite strengths and encompasses a wide array of failure mechanisms, including both GLS and Local Load Sharing (LLS) models.

4.1.1 A Conventional Model

The conventional GLS model which is presented by Curtin [2] will be mentioned briefly here. The model estimated the rupture strain and stress of unidirectional composites considering the fiber failure probability by assuming that each fiber fails independently. The interfacial shear stress was assumed to be constant in the region of the stress recovery length (L_r), as shown in Figure 4.1, which shows fiber-axial stress distribution around a broken fiber. The average of fiber stress $\sigma_{f,\,ave}$ is

$$\sigma_{f,\,ave} = \left(1-q\right)E_f\varepsilon + \frac{qE_f\varepsilon}{2}$$ Eq. (4.1)

where ε is the composite strain, E_f is the elastic modulus of fiber, q is the probability of fiber failure in the $2L_r$ length, and the fiber is assumed to be elastic. This equation is consistent when ε is small. Generally, using the Weibull

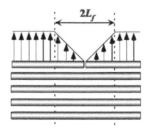

FIGURE 4.1   Axial fiber stress distribution around a broken fiber.

parameters L_0, m, and σ_0, the fiber failure probability is as follows:

$$P = 1 - \exp\left(-\frac{L}{L_0}\left(\frac{\sigma}{\sigma_0}\right)^m\right)$$

Eq. (4.2)

The probability of fiber break q in the $2L_r$ length is

$$q = 1 - \exp\left(-\frac{2L_r}{L_0}\left(\frac{E_f\varepsilon}{\sigma_0}\right)^m\right)$$

Eq. (4.3)

Here $E_f\varepsilon$ can be substituted for the fiber stress σ. From the force balance equation (well known as Kelly–Tyson equation [3]), L_r can be derived using the interfacial shear stress (τ_0) that can correspond to the interfacial shear strength and fiber diameter (D) as follows:

$$L_r = \frac{E_f\varepsilon D}{4\tau_0}$$

Eq. (4.4)

Eventually, Eq. (4.1) can be described as

$$\bar{\sigma}_f = E_f\varepsilon\left\{1 - \frac{1}{2}\left(\frac{E_f\varepsilon}{S_c}\right)^{m+1}\right\}$$

Eq. (4.5)

where $S_c = \left(\dfrac{2\sigma_0{}^m\tau_0 L_0}{D}\right)^{1/(m+1)}$.

Here, an approximation $q = \left(\dfrac{2L_r}{L_0}\right)\left(\dfrac{E_f\varepsilon}{\sigma_0}\right)^m$ is used on an assumption of ε is small. $\dfrac{\partial\bar{\sigma}_f}{\partial\varepsilon} = 0$ leads to the critical strain (ε_{cri}), and then the maximum fiber average stress (S_{max}) can be obtained by the critical strain being substituted into Eq. (4.5). These results are described as follows:

$$S_{max} = S_c\frac{m+1}{m+2}\left(\frac{2}{m+2}\right)^{1/(m+1)}$$

Eq. (4.6)

$$\varepsilon_{cri} = \frac{S_c}{E_f}\left(\frac{2}{m+2}\right)^{1/(m+1)}$$

Eq. (4.7)

The critical strain corresponds to the rupture strain, and the maximum value, multiplied by the fiber volume fraction, corresponds to the rupture stress of a unidirectional composite when we assume that all of the load is applied to only the fiber. This model has been verified to be applicable to some types of composites [2–4] but not for a wide range of composites.

4.1.2 SFF Model [1]

The gap between strengths obtained from experiments and estimations is attributed to the fact as follows: in the model, fiber is assumed to fail completely individually. It means all fibers are supposed to work properly. But in actuality, in experiments, some of fibers fail simultaneously as a group. It means that even intact and strong fiber fails due to neighboring failure of weak fiber so that total load capacity of fiber bundle is less than that of all fibers in total. This phenomenon is validated by acoustic emission experiments and fractography investigations [1].

Here, we will present a new model that can represent the type of complicated relationships mentioned earlier. We deal first with the number of fibers in a simultaneous failure group as an unknown parameter. Based on the Weibull statistics, fiber failure probability q' in a fiber group consisting of n number of fibers in a modified stress recovery length L_r' can be expressed as given in the following equation,

$$q' = 1 - \exp\left(-\frac{2nL_r'}{L_0}\left(\frac{E_f \varepsilon}{\sigma_0}\right)^m\right)$$

Eq. (4.8)

Here, L_r' is the function of the elastic modulus and diameter of a simultaneous failure fiber group composite, respectively. This equation is derived by applying the weakest-link theory to the fiber-group failure probability similar to Eq. (4.3). In other words, Eq. (4.8) is based on the fact that the weakest "element" in $n \times 2L_r'$ governs the fiber-group strength. The values of the elastic modulus and diameter vary with a change in n. On a consideration for the force balance of the simultaneous failure fiber-group composite consisting of n numbers of fiber and their accompanying matrix, the modified stress recovery length L_r' can be derived as

$$L_r' = \frac{E_f \varepsilon D}{4\tau_0} \cdot n\sqrt{\frac{V_f}{V_f + n - 1}}$$

Eq. (4.9)

$$\cong \frac{E_f \varepsilon D}{4\tau_0} \sqrt{V_f n} \quad \text{(when } 1 \ll n).$$

Equation (4.9) satisfies $L_r' = L_r$ when $n = 1$. As in the conventional model, the critical strain and maximum fiber average stress considering the fiber group unit fracture are derived as follows:

$$\varepsilon_{cri}' = \frac{S_c}{E_f} \left(\frac{2}{m+2} \cdot \frac{1}{n^2} \sqrt{\frac{V_f + n - 1}{V_f}} \right)^{1/(m+1)} \qquad \text{Eq. (4.10)}$$

$$\cong \frac{S_c}{E_f} \left(\frac{2}{m+2} \cdot \frac{1}{\sqrt{V_f \cdot n^3}} \right)^{1/(m+1)} \quad \text{(when } 1 \ll n)$$

$$S_{max}' = S_c \frac{m+1}{m+2} \left(\frac{2}{m+2} \cdot \frac{1}{n^2} \sqrt{\frac{V_f + n - 1}{V_f}} \right)^{1/(m+1)} \qquad \text{Eq. (4.11)}$$

$$\cong S_c \frac{m+1}{m+2} \left(\frac{2}{m+2} \cdot \frac{1}{\sqrt{V_f \cdot n^3}} \right)^{1/(m+1)} \quad \text{(when } 1 \ll n)$$

We refer to this model as Simultaneous Fiber-Failure (SFF) model. If we then assume S_{max}' corresponds to the maximum fiber average stress derived from the experimental result considering the fiber volume fraction ($S_{max}' = S_{exp} = \sigma_{UD}/V_f$), we can obtain the number of simultaneous fiber failures n from Eq. (4.11) divided by Eq. (4.6) as follows:

$$\frac{S_{max}'}{S_{max}} = \left(\frac{1}{n^2} \sqrt{\frac{V_f + n - 1}{V_f}} \right)^{1/(m+1)} \qquad \text{Eq. (4.12)}$$

Here, we describe how the value of n, the number of fibers that fail simultaneously, is determined. A key point to consider is whether a fiber failure causes a neighboring fiber to fail. Kendall has studied crack propagation around a bi-material interface. The condition in which a crack in material 2 is deflected along the interface is governed by

$$\frac{G_{ic}}{G_{2c}} \leq \left(\frac{h_1 E_1 + h_2 E_2}{h_2 E_2} \right) \left[\frac{1}{4\pi(1 - v^2)} \right], \qquad \text{Eq. (4.13)}$$

where G_c, E, v, and h are the critical energy release rate, Young's modulus, Poisson's ratio of the fiber and matrix, and the thickness, respectively. The subscripts 1, 2, and i in Eq. (4.13) represent the properties of material 1, material 2, and the interface, respectively. In other words, whether the material fails or the interface fails when the neighboring material fails is determined by the toughness ratio of the material and interface (G_{ic}/G_{2c}). Note that this model is two-dimensional under a tensile load.

This approach is adapted to suit our research, as illustrated in Figure 4.2. The failure of the right-side fiber when the left-side fiber fails is determined by two conditions. First, the left-side fiber failure should lead to a matrix failure through the interface, which exhibits higher interface toughness compared to the matrix toughness (Gi/Gm); second, the matrix failure should subsequently cause the right-side fiber to fail through the interface, which exhibits toughness relative to the fiber toughness (Gi/Gf). It can be anticipated that the neighboring fiber fails if these two conditions are satisfied when a fiber fails. The fiber failures spread until the crack comes into contact with a weak interface. This is the mechanism that generates a simultaneous fiber-failure group. In the present study, we employ strengths instead of toughness (τ_i/σ_m and τ_i/σ_0 instead of G_i/G_m and G_i/G_f, respectively). Moreover, the geometrical average of the two strength ratios ($\tau_i / \sqrt{\sigma_0\sigma_m}$) is assumed to be an index that governs the number of the simultaneous fiber-failures group (n). Incorporating strength in this study presents a challenge. Strength and toughness are distinct concepts. However, the strength ratio might also be able to govern the phenomenon of whether a neighboring fiber fails or not, as depicted in Figure 4.2. It can be predicted that if the interface is relatively strong, both the matrix and neighboring fiber are more likely to fail. This prediction is at least reasonably consistent with the

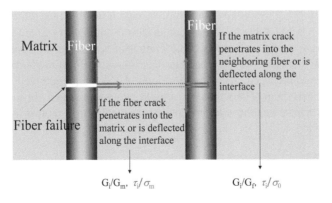

FIGURE 4.2 Fiber failure propagation.

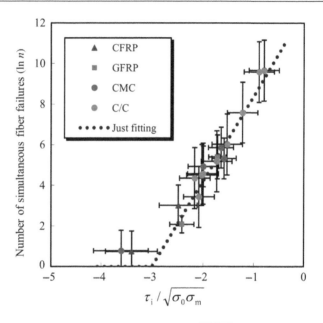

FIGURE 4.3   Relationship between n and $\tau_i / \sqrt{\sigma_0 \sigma_m}$ in various composites.

failure-surface observations [1], as previously discussed. Additionally, using strength instead of toughness simplifies the model and facilitates its application due to the relative ease of data acquisition. The matrix strength is the only additional data required, aside from the conventional GLS data. Owing to these factors, we attempt to establish an empirical relationship between n and the strength ratio, even though employing the toughness ratio might be more appropriate.

Figure 4.3 shows the relationship between n obtained by Eq. (4.12) and $\tau_i / \sqrt{\sigma_0 \sigma_m}$ using Table 4.1 for all composites examined in this study. In this figure, each error bar shows a consideration of scattering the experimental results (including the deviation due to the difference of test methods) and the experimental precision. As shown in Figure 4.3, a unique interaction can be seen between them. The value of n should be no less than 1. In the present study, we determined the relationship between them with the following empirical equation:

$$\ln(n) = \alpha \cdot \ln\left(\tau_i / \sqrt{\sigma_m \sigma_0}\right) + \beta \ (n > 1) \qquad \text{Eq. (4.14)}$$

where $\alpha = 4.2$ and $\beta = 12.6$, both of which are determined by "Just fitting" in Figure 4.3.

TABLE 4.1 Materials and Experimental Results

NO.	MATERIAL	σ_{UD} (MPa)	V_F	σ_0 (MPa)	M	τ_i (MPa)	σ_M (MPa)
1	E-glass/phenol resin	640	0.4	2,500	13	40	20
2	E-glass/vinyl ester	252	0.1	2,500	13	40	80
3	carbon (M40)/vinyl ester	1390	0.3	4,500	16	20	80
4	Carbon (M40)/ epoxy	2310	0.6	4,500	16	50	80
5	C/C (1,000°C heat treated)	410	0.6	3,100	6.5	56	5
6	C/C (1,300°C heat treated)	400	0.6	2,900	6.7	49	5
7	C/C (1,600°C heat treated)	520	0.6	3,000	5.9	36	5
8	C/C (2,000°C heat treated)	762	0.6	3,100	6.4	27	5
9	C/C (2,300°C heat treated)	798	0.6	3,100	5.4	22	5
10	C/C (2,600°C heat treated)	852	0.6	3,150	4.9	17	5
11	C/C (2,900°C heat treated)	1,020	0.6	3,400	5.8	15	5
12	C/C (resin char method)	1,360	0.6	3,700	5.7	17	5
13	C/C (hot isostatic pressing method)	1,080	0.6	3,700	5.7	18	5
14	CMC (Curtin's value in ref [1])	700	0.5	1,700	4	5	20
15	CMC (thick-coated interface/ conventional PIP matrix)	250	0.2	3,600	5	20	3
16	CMC (thick-coated interface/ modified PIP matrix)	320	0.2	3,600	5	20	6
17	CMC (thin-coated interface/ modified PIP matrix)	310	0.2	3,600	5	30	6

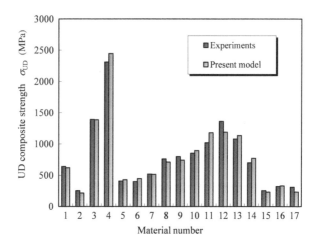

FIGURE 4.4   Comparison of predicted strength with experimental results.

Figure 4.4 shows a comparison between the experimental results and the predicted strength obtained from Eq. (4.14) and Eq. (4.11) × V_f. As shown in this figure, the predictions are improved drastically. The validity of the SFF is indicated by these results. In addition to this, the SFF can give close values of the tensile strength in other articles with practical assumptions for the properties which is not mentioned in them. For example, for $\sigma_0 = 5,700$ MPa, $m = 10$, $L_0 = 20$ mm (assumption), $D = 0.006$ mm, $\sigma_m = 80$ MPa (assumption of normal epoxy resin strength), $\tau_i = 70$ MPa (assumption of the carbon/epoxy interface in Ref. [5]), $V_f = 0.6$ for T800H/#3633 unidirectional composite, the SFF gives 2,670 MPa and Ref. [6] shows 2,700 MPa. And, for $\sigma_0 = 3,700$ MPa, $m = 5.7$ (same as in this study), $L_0 = 25$ mm, $D = 0.006$ mm, $\sigma_m = 6$ MPa (1.5 times of transverse strength of UD [7]), $\tau_i = 50$ MPa [7], $V_f = 0.28$ (longitudinal V_f) for C/C under high temperature in Ref. [7], the SFF gives 273 MPa and Ref. [7] shows 250 MPa.

4.1.3 Discussion

We will mention here the applicability of the SFF on the assumption that the mechanical properties of fiber and fiber volume fraction are assumed to be constant ($L_0 = 20$ mm, $\sigma_0 = 4,900$ MPa, $m = 10$, $D = 0.006$ mm, and $V_f = 0.6$). Figure 4.5(a) and (b) show the unidirectional composite strength as a function of the interfacial strength and as a function of the matrix strength, respectively. In Figure 4.5(a), the unidirectional composite strength is enhanced first with

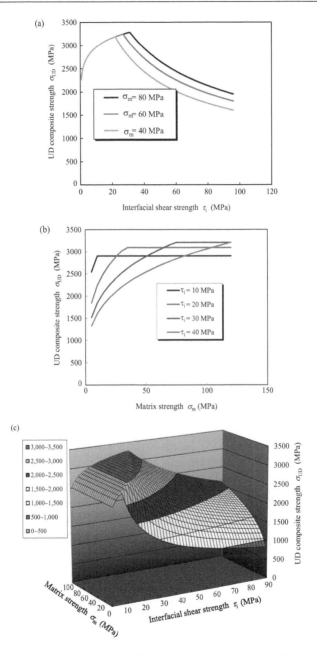

FIGURE 4.5 Analytical results: unidirectional composite strength (a) as a function of interfacial shear strength, (b) as a function of matrix strength, and (c) as functions of interfacial shear strength and matrix strength.

an increase in the interfacial strength, and then it decreases with an increase in the interfacial strength. The position of the inflection point is dominated by the matrix strength. On the left side of the point, it corresponds to the conventional GLS model, and the fibers fail independently ($n = 1$). On the right side of the point, the fibers fail simultaneously, and the composite fails in a brittle manner. The SFF can lead to relationships between the composite strength and interfacial strength exactly similar to those of experimental [5, 6] and analytical results [8, 9] in the existing articles.

In Figure 4.5(b), the composite strength does not vary with the matrix strength in the flat regions. There, the predicted unidirectional composite strength corresponds to the conventional model and $n = 1$. That is why the matrix strength does not matter for Curtin's data, which is in the regions, as mentioned earlier. However, if the matrix strength is less than specific values, the composite strength decreases with a decrease in the matrix strength. The specific values depend on the interfacial strength. As shown in Table 4.1, the composite strengths of specimens no. 1 and 2 are widely different on an assumption of identical fiber volume fraction, and they are different only in matrix strength. This fact can be explained easily by Figure 4.5(b). The phenolic matrix is weaker than the vinyl ester matrix. Furthermore, C/C composite under high temperature is generally stronger than that under room temperature. This can be also explained by Figure 4.5(b) and the fact that the transverse strength which corresponds to matrix strength in this case under high temperature is stronger than that under room temperature [7]. Finally, Figure 4.5(c) summarizes the unidirectional composite strength as functions of the interfacial strength and the matrix strength in a 3D view graph. Thus, SFF is good enough to predict and estimate tensile failure strength of UD-CFRP.

4.2 MICROMECHANICAL NUMERICAL SIMULATIONS FOR FIBER-AXIAL SHEAR ANALYSIS

In this chapter, we perform 3D PUC analysis to obtain the shear properties of composite materials. In this study, a three-dimensional periodic unit cell (3D PUC) is created using FEA software Abaqus 2019/Implicit, which is shown in Figure 4.6. The finite element model consists of epoxy resin and carbon fibers, and the interface between the fibers and the resin is not considered in this study because it is considered strong enough from the study by Sato M. et al. [10].

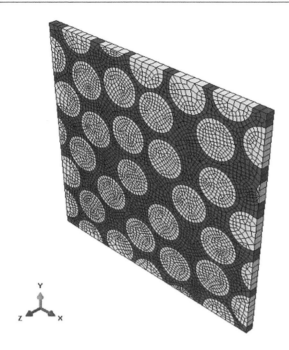

FIGURE 4.6 3D PUC model.

TABLE 4.2 Mechanical Properties of Fibers

E_1	E_2	E_3	N_{12}	N_{13}	N_{23}	G_{12}	G_{13}	G_{23}
14 GPa	14 GPa	294 GPa	0.35	0.02	0.02	5 GPa	18 GPa	18 GPa

The model contained 29 fibers with a fiber diameter of 6 μm. Both the height and width of the model are about 39 μm, and the volume fraction of fibers is 55%. Eight node elements (C3D15) are used in the model. The number of elements is 38,322, and the number of nodes is 173,760. Matrix initial modulus is 3.6 GPa, and Poisson's ratio is 0.34. The material properties of fiber is shown in Table 4.2. The resin is isotropic and elastoplastic material, considering continuous damage mechanics. Fibers are modeled as orthotropic elastic materials.

A periodic boundary condition is applied to each plane. Key degrees of freedom method [11] is applied in this simulation. This method is very popular because of its easiness of usage so that it has been applied in many studies [10, 12–14]. Employing this method, pure shear deformation is applied in this model.

In this study, the stress–strain relationship and the damage initiation criterion of resin proposed by Yamazaki et al. [15] were used. The stress–strain relationship of the resin was obtained from the static compressive test of the polyimide coupon specimen. The relationship between the equivalent stress and equivalent strain in the resin was extrapolated in order to comply with the stress–strain relationship based on the compressive static test. The resin's damage initiation stress was determined from the experimental results and by Eqs (4.15) and (4.16).

$$\frac{T}{C} = \frac{2}{5} \qquad \text{Eq. (4.15)}$$

$$\left(1 - \frac{T}{C}\right)\left(\sigma_1 + \sigma_2 + \sigma_3\right) + \frac{1}{2TC}\left[\left(\sigma_1 - \sigma_2\right)^2 + \left(\sigma_2 - \sigma_3\right)^2 + \left(\sigma_3 - \sigma_1\right)^2\right] \le 1$$

$$\text{Eq. (4.16)}$$

where T and C are the uniaxial tensile strength and uniaxial compressive strength, respectively. Eq. (4.16) is the failure criterion proposed by Christensen [16, 17]. Resin's isotropic Christensen failure envelope obtained by Eqs (4.15) and (4.16) is shown in Figure 4.7.

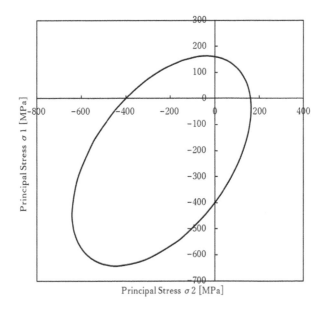

FIGURE 4.7   Resin's isotropic Christensen failure envelope obtained by Eqs (4.15) and (4.16).

Table 4.3 shows the results of the 3D PUC in-plane shear analysis. From the shear stress–strain relationship, the elastic and plastic in-plane shear modulus, G_{12}^{ε} and G_{12}^{p}, the yield shear stress, τ_{y}, and the shear failure stress, τ_{ult}, are read, and the values are shown in the table. These four parameters are utilized in subsequence section, which is an estimation of fiber-axial compressive strength.

4.3 FIBER-AXIAL COMPRESSIVE STRENGTH

We determined fiber-axial compressive strength from the fiber microbuckling model of Berbinau et al. [18, 19]. The Berbinau fiber microbuckling model was based on the initial sinusoidal shape of the fiber. Berbinau et al. modelled the initial fiber waviness using sine function $v_0(x)$.

$$v_0(x) = V_0 \sin\left(\frac{\pi x}{\lambda_0}\right)$$ Eq. (4.17)

where V_0 is the amplitude of the initial fiber waviness and λ_0 is its half wavelength. When a compressive load was applied, the fiber deformed into new sine function $v(x)$.

$$v(x) = V \sin\left(\frac{\pi x}{\lambda}\right)$$ Eq. (4.18)

where V is the amplitude of the fiber waviness and λ is its half wavelength. The above fiber shapes and the balance state at the fiber microlength are shown in Figure 3.5–3.6.

From Figure 4.8, the equilibrium of force and moment are given by Eq. (4.19) and (4.20), respectively.

TABLE 4.3 The In-plane Shear Properties of CFRP Obtained From the 3D PUC Analysis

Elastic shear modulus	6,530 MPa
Plastic shear modulus	300 MPa
In-plane shear yield stress	85 MPa
In-plane shear strength	140 MPa

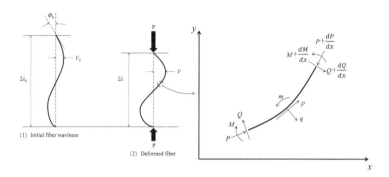

FIGURE 4.8   The fiber shapes and the balance state at the fiber microlength.

$$x;\, p - \frac{dP}{dx} + \frac{dQ}{dx}\frac{dv}{dx} = 0$$

$$y;\, q - \frac{dQ}{dx} + \frac{dP}{dx}\frac{dv}{dx} = 0 \qquad \text{Eq. (4.19)}$$

$$moment;\, \frac{dM}{dx} - Q + m = 0 \qquad \text{Eq. (4.20)}$$

where P is the applied compressive load and Q is the shear force and M is the moment.

If we assume that $p = q = 0$ since all fibers in CFRP deform in the same way, Eqs (4.19) and (34.20) can be organized into Eq. (4.21).

$$\frac{d^2 M}{dx^2} + P\frac{d^2 v}{dx^2} + \frac{dm}{dx} = 0 \qquad \text{Eq. (4.21)}$$

Eqs (4.22)–(4.25) are the equations of the deflection curve, of shear force and deformation, of stress on the fiber, and of shear stiffness.

$$\frac{d^2(v - v_0)}{dx^2} = \frac{M}{E_f I} \qquad \text{Eq. (4.22)}$$

$$\frac{dm}{dx} = -A_f G_{12}^{ep}(\gamma)\frac{d^2(v - v_0)}{dx^2} \qquad \text{Eq. (4.23)}$$

$$\sigma_{fibre\,microbuckling} = \sigma_0 = \frac{PV_f}{A_f} \qquad \text{Eq. (4.24)}$$

$$G_{12}^{ep}(\gamma) = G_{12}^{e} exp\left(-\frac{G_{12}^{e}\gamma}{\tau_{y}}\right) + G_{12}^{p} exp\left(-\frac{G_{12}^{p}\gamma}{\tau_{ult} - \tau_{y}}\right)$$ Eq. (4.25)

where I is the moment of inertia of area, A_f is the fiber cross section, G_{12}^{ep} is the composite shear modulus, G_{12} and G_{12}^{p} are the elastic and plastic in-plane shear modulus, respectively, τ_y is the yield shear stress, γ is the shear strain, and τ_{ult} is the shear failure stress.

By substituting Eqs (4.22)–(4.25) into Eq. (4.21), Eq. (4.21) can be expressed as Eq. (4.26).

$$E_f I \frac{d^4(v-v_0)}{dx^4} + \frac{A_f \sigma_0}{V_f} \frac{d^2 v}{dx^2} - A_f G_{12}^{ep}(\gamma) \frac{d^2(v-v_0)}{dx^2} = 0$$ Eq. (4.26)

If $\lambda \approx \lambda_0$, Eq. (4.26) can be transformed as Eq. (4.27).

$$\frac{V}{V_0} = \left[\left(1 - \frac{P}{E_f I(\pi/\lambda)^2 - A_f G}\right)^{-1}\right]$$ Eq. (4.27)

V/V_0 is written in the form of Eq. (4.27). V/V_0 slowly increases with the increase in the applied stress σ and then exponentially grows until it reaches maximum stress. Equation (4.27) assumes that V in the function increases rapidly and that the fiber buckles at the point of the asymptote. The stress at the point of asymptote is defined as the compressive strength. In other words, this compressive strength prediction model uses the fiber buckling to predict the compressive strength.

As can be seen from Eq. (4.27), the composite shear mechanical properties are required to predict the compressive strength using the microbuckling model. Therefore, we obtained the shear properties from 3D PUC analysis in this study. Using the shear properties obtained from the 3D PUC analysis, from Eq. (4.27), the expected compressive strength is about 1,600 MPa.

4.4 DISCUSSION

In this chapter, micromechanical models of fiber-axial tensile, shear, and compressive strengths are introduced. Here, transverse tensile, compressive, and shear strengths of CFRP are not presented. But these can be readily implemented by PUC simulation as shown in Figure 4.6. Also, if we want to introduce various failure criteria for resin matrix such as entropy-based failure criteria, it is not difficult to implement into the numerical simulation like Figure 4.6. Some of those examples are introduced later in this book. The same situation can be applied to fiber-axial strengths. It is very easy to apply preferable resin failure criterion, which can be employed to SFF model and microbuckling model, too. Thus, we can simulate many kinds of CFRP failures using these failure criteria.

REFERENCES

[1] J. Koyanagi, H. Hatta, M. Kotani, H. Kawada, A comprehensive model for determining tensile strengths of various unidirectional composites, Journal of Composite Materials 43(18) (2009) 1901–1914.

[2] W.A. Curtin, Theory of mechanical properties of ceramic-matrix composites, Journal of the American Ceramic Society 74(11) (1991) 2837–2845.

[3] J. Koyanagi, G. Kiyota, T. Kamiya, H. Kawada, Prediction of creep rupture in unidirectional composite: Creep rupture model with interfacial debonding and its propagation, Advanced Composite Materials: The Official Journal of the Japan Society of Composite Materials 13(3–4) (2004) 199–213.

[4] J. Koyanagi, H. Hatta, F. Ogawa, H. Kawada, Time-dependent reduction of tensile strength caused by interfacial degradation under constant strain duration in UD-CFRP, Journal of Composite Materials 41(25) (2007) 3007–3026.

[5] S. Subramanian, K.L. Reifsnider, W.W. Stinchcomb, Tensile strength of unidirectional composites: The role of efficiency and strength of fiber-matrix interface, Journal of Composites Technology and Research 17(4) (1995) 289–300.

[6] JAXA, Advanced Composites Database System: JAXA-ACDB; Ver. 06–1. 2022.

[7] K. Goto, H. Hatta, M. Oe, T. Koizumi, Tensile strength and deformation of a two-dimensional carbon-carbon composite at elevated temperatures, Journal of the American Ceramic Society 86(12) (2003) 2129–2135.

[8] K. Goda, The role of interfacial debonding in increasing the strength and reliability of unidirectional fibrous composites, Composites Science and Technology 59(12) (1999) 1871–1879.

[9] H. Li, Y. Jia, G. Mamtimin, W. Jiang, L. An, Stress transfer and damage evolution simulations of fiber-reinforced polymer-matrix composites, Materials Science and Engineering A 425(1–2) (2006) 178–184.

[10] M. Sato, S. Shirai, J. Koyanagi, Y. Ishida, Y. Kogo, Numerical simulation for strain rate and temperature dependence of transverse tensile failure of unidirectional carbon fiber-reinforced plastics, Journal of Composite Materials 53(28–30) (2019) 4305–4312.

[11] K. Terada, N. Hirayama, K. Yamamoto, S. Matsubara, Numerical plate testing for linear multi-scale analyses of composite plates, Transactions of the Japan Society for Computational Engineering and Science 2015 (2015) 20150001–20150001.

[12] J. Koyanagi, Y. Sato, T. Sasayama, T. Okabe, S. Yoneyama, Numerical simulation of strain-rate dependent transition of transverse tensile failure mode in fiber-reinforced composites, Composites Part A: Applied Science and Manufacturing 56 (2014) 136–142.

[13] T. Okabe, H. Imamura, Y. Sato, R. Higuchi, J. Koyanagi, R. Talreja, Experimental and numerical studies of initial cracking in CFRP cross-ply laminates, Composites Part A: Applied Science and Manufacturing 68 (2015) 81–89.

[14] M. Sato, K. Hasegawa, J. Koyanagi, R. Higuchi, Y. Ishida, Residual strength prediction for unidirectional CFRP using a nonlinear viscoelastic constitutive equation considering entropy damage, Composites Part A: Applied Science and Manufacturing 141 (2021).

[15] Y. Yamazaki, J. Koyanagi, Y. Sawamura, M. Ridha, S. Yoneyama, T.E. Tay, Numerical simulation of dynamic failure behavior for cylindrical carbon fiber reinforced polymer, Composite Structures 203 (2018) 934–942.

[16] R.M. Christensen, Failure criteria for fiber composite materials, the astonishing sixty year search, definitive usable results, Composites Science and Technology 182 (2019) 107718.

[17] R. Kitamura, T. Kageyama, J. Koyanagi, S. Ogihara, Estimation of biaxial tensile and compression behavior of polypropylene using molecular dynamics simulation, Advanced Composite Materials 28(2) (2019) 135–146.

[18] P. Berbinau, C. Soutis, I.A. Guz, Compressive failure of $0°$ unidirectional carbon-fibre-reinforced plastic (CFRP) laminates by fibre microbuckling, Composites Science and Technology 59(9) (1999) 1451–1455.

[19] A. Jumahat, C. Soutis, F.R. Jones, A. Hodzic, Fracture mechanisms and failure analysis of carbon fibre/toughened epoxy composites subjected to compressive loading, Composite Structures 92(2) (2010) 295–305.

Durability Predicted by Microscale Simulations

5

5.1 TIME AND TEMPERATURE DEPENDENCE OF TRANSVERSE TENSILE FAILURE OF UNIDIRECTIONAL CARBON-FIBER-REINFORCED POLYMER MATRIX COMPOSITES

5.1.1 Experimental Works [1]

Carbon-fiber-reinforced plastics (CFRPs) have gained widespread use in industries such as aerospace and automotive vehicles. Precise failure prediction of CFRP is essential for ensuring safety. While static tensile failure is often the primary concern, multiple failure modes are associated with composite failure. Generally, the ultimate failure of CFRP laminates is governed by fiber-directional lamina strength [2–4]. However, the inception of cracks is determined by transverse lamina strength, commonly known as "transverse cracks". Although transverse cracks may not significantly affect ultimate strength, they can cause leakage issues and initiate interlaminar debonding. Transverse crack strength is primarily determined by matrix strength, which is time and temperature dependent. Consequently, the strength of transverse cracks is also time and temperature dependent. This chapter delves into the

DOI: 10.1201/9781003371137-5

time and temperature dependence of transverse strength in unidirectional CFRPs using both experimental and numerical approaches.

Initially, a concise summary of one of the authors' experimental works [1] is provided. Figure 2.6 displays SEM fractographs of failure surfaces for transverse tensile testing of unidirectional CFRP. These tests were conducted at varying temperature and strain rates. The failure morphology can be categorized into two groups. One group exhibits interface failure, with fibers visible on the surface. The other group demonstrates matrix failure, where the fiber is barely discernible on the surface. As such, the former group is dominated by interface failure, while the latter is dominated by matrix failure.

The transition between these failure modes occurs because the strain rate and temperature dependence of interface strength differ from that of matrix strength. If both strengths were equivalent, the failure mode transition would not take place. Comprehensive failure is instigated by microfailure, where stress concentration reaches its peak, referred to as the "critical point". Microfailure, either interface or matrix failure, transpires near the critical point. To comprehend and predict this failure mode transition for transverse tensile failure in unidirectional composites, knowledge of the strain rate and temperature dependence of both interface strength and matrix strength is necessary.

5.1.2 Simple FEA and Critical Point Stress [1]

This chapter touches upon the strain rate and temperature dependence of interface strength. By assuming a square fiber alignment, the simplest unit cell serves as the representative volume element. The critical point is situated at the upper edge of the fiber–matrix interface. At this location, either the interface or the matrix near the critical point experiences failure first, followed by the development of complete failure. The occurrence of such failure is determined by whether the interface strength or matrix strength is weaker. The critical point of stress at specimen failure can be obtained by

$$\sigma_{cp} = K \cdot \sigma_{ult} - \sigma_r \quad (1) \qquad\qquad \text{Eq. (5.1)}$$

The critical-point stress at specimen failure can be obtained using Eq. (5.1), where σ_{cp} is the microstress at the critical point, K is the stress concentration factor, σ_{ult} is the specimen's stress at ultimate failure, and σ_r is the thermal residual compressive stress at specimen failure. Both the stress concentration factor and thermal residual stress depend on time and temperature. To accommodate this, a time–temperature superposition principle is employed in this study in which elevated temperature is considered as time acceleration, or "reduced time". The time acceleration is determined using Arrhenius' plot.

The stress concentration factor and residual stress at the critical point, as functions of reduced time, are numerically analyzed, and the critical-point stress can be calculated using Eq. (5.1), and the result is shown in Figure 2.7.

The findings can be divided into two groups. The first group is characterized by a constant critical-point strength, indicating that it is not dominated by reduced strain rate, while the second group demonstrates a decline in critical-point strength as the reduced strain rate decreases. Comparing these results with the failure morphology discussed in the earlier section, the former corresponds to the interface-failure-dominant mode, and the latter corresponds to the matrix-dominant mode. This implies that the interface strength is not dependent on time (strain rate) and temperature.

5.1.3 Time and Temperature Dependence on Interface Strength for Microscale Numerical Simulation

The time and temperature dependence of interface strength is a topic of ongoing research and discussion. While some studies have indicated that interface strength is not time or temperature dependent [5, 6], others have suggested that it may be [2, 7, 8]. To clarify this issue, it is important to determine the actual interface stress through finite-element analysis or another suitable method. Once the interface stress is known, the choice of failure criterion can have a significant impact on the results. In particular, a parabolic criterion for interface failure envelope, as shown in Figure 2.1, may provide more accurate results than the commonly used quadratic criterion in numerical analyses.

5.1.4 RVE Model

The representative volume element (RVE) approach is a widely recognized and highly effective method for modeling CFRP, with numerous studies available in the literature [9–19]. These studies have investigated various aspects related to RVE modeling, such as the necessary number of fibers, effects of fiber alignment, and other pertinent factors. Figure 5.1 presents an example of an RVE, which is a periodic unit cell composed of 30 fibers and surrounding matrix resin. In this section, we offer a concise overview of the numerical procedure employed for simulating the behavior of unidirectional CFRP under transverse tensile loading. The RVE is defined with a fiber diameter of 6 μm and a cell size of 39 × 39 μm, resulting in a fiber volume fraction of 56%. The fiber is assumed to be an elastic body, while the matrix is considered an elastoplastic body involving continuum damage mechanics (CDM). Both materials are assumed to be isotropic in the plane of analysis. The elastic modulus for the fiber and matrix is set at 14 GPa and 3 GPa, respectively, with Poisson's ratio values of 0.4 and 0.34. The linear thermal expansion coefficients are assigned at 1E-05 and 6E-05 for fiber and matrix, respectively. To simulate interfacial debonding, a cohesive zone modeling (CZM) technique is applied to the fiber/matrix interface.

The analysis comprises two steps: thermal stress analysis and tensile failure simulation. To assess the temperature dependence of composite strength, it is crucial to consider the difference in local thermal stress. The curing temperature is assumed to be 200°C, while the test temperatures are set at −50°C, 0°C, and 50°C. Strain rates of 10^{-6}, 10^{-4}, 0.01, and 1 s^{-1} are chosen for each temperature. Thermal residual stress and the change in matrix strength due to temperature are two factors that influence the tensile behavior of the composite. Generally, matrix strength decreases with increasing temperature, but the effect of strain rate also impacts matrix strength. To accommodate this, a time–temperature superposition principle is employed, in which the difference in temperature is considered as the difference in strain rate for defining matrix strength. The elastic modulus is assumed to be independent of temperature and strain rate, which may not accurately represent the actual phenomenon. However, when discussing strength alone, the assumption of a constant modulus has a minimal impact. The critical point, where complete failure develops, is identified on the basis of the numerical analysis. Fiber failure is not considered in this simulation, and a plane stress condition is assumed. The simulation of unidirectional CFRP under transverse tensile loading is conducted by applying periodic boundary conditions, CDM, and CZM. These procedures are described in detail in the following sections.

5.1.5 Periodic Boundary Conditions

This study employs the method suggested by Li et al. [20] for periodic boundary conditions, which is known as the key degree of freedom (DOF) method. This method places an independent node out-of-model, and the relative displacement of the pair of nodes in the periodic boundary condition is related to the displacement of the key DOF of the node. This enables the control of global strain and observation of global stress readily. The key DOF method has recently been used in numerical modeling of homogenization [21–23].

Figure 5.1 shows a unit cell and the definitions of periodic boundary edges and points: edges (A)–(D) and points (a)–(d). Note that the points are not included in any edges. The length of the unit cell is defined as L.

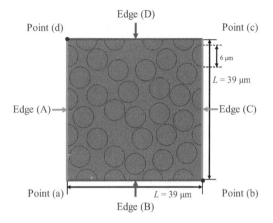

FIGURE 5.1   Example of periodic unit cell consisting of 30 fibers and surrounding matrix and definitions of edges and points.

The displacement u_i of arbitrary position x_i of the internal unit cell is described using global strain E_{ij} and microscopic perturbed displacement $u_i^{(per)}$ as follows:

$$\begin{cases} u_1(x_1,x_2) = E_{11}x_1 + E_{12}x_2 + u_1^{(per)}(x_1,x_2) \\ u_2(x_1,x_2) = E_{12}x_1 + E_{22}x_2 + u_2^{(per)}(x_1,x_2) \end{cases}$$ Eq. (5.2)

The periodic boundary condition between Edges (A) and (C) is considered here. Microscopic displacements of corresponding nodes on Edges (A) and (C), $u_i^{(per)}\big|_A$ and $u_i^{(per)}\big|_C$, should be the same because of periodicity; thus,

$$u_i^{(per)}\big|_A = u_i^{(per)}\big|_C$$ Eq. (5.3)

When Eq. (5.3) is substituted into Eq. (5.2), they give:

$$\begin{cases} u_1\big|_C - u_1\big|_A = LE_{11} \\ u_2\big|_C - u_2\big|_A = LE_{12} \end{cases}$$ Eq. (5.4)

Eq. (5.4) presents boundary conditions that should be applied between Edges (A) and (C) in this study. The boundary conditions between Edges (B) and (D) are determined as well:

$$\begin{cases} u_1\big|_D - u_1\big|_B = LE_{12} \\ u_2\big|_D - u_2\big|_B = LE_{22} \end{cases}$$ Eq. (5.5)

To eliminate duplicate boundary conditions, the following boundary conditions are applied to points besides those in Eqs (5.4) and (5.5):

Between Point (a) and Point (b),

$$\begin{cases} u_1\big|_b - u_1\big|_a = LE_{11} \\ u_2\big|_b - u_2\big|_a = LE_{12} \end{cases}$$

Eq. (5.6)

Between Point (a) and Point (c),

$$\begin{cases} u_1\big|_c - u_1\big|_a = LE_{11} + LE_{12} \\ u_2\big|_c - u_2\big|_a = LE_{12} + LE_{22} \end{cases}$$

Eq. (5.7)

Between Point (a) and Point (d),

$$\begin{cases} u_1\big|_d - u_1\big|_a = LE_{12} \\ u_2\big|_d - u_2\big|_a = LE_{22} \end{cases}$$

Eq. (5.8)

In addition to the above conditions, $u_1 = u_2 = 0$ is applied to an arbitrary node to eliminate a rigid body displacement.

Next, we introduce the key node method for periodic boundary conditions. In the present study, Eqs (5.4)–(5.8) involve the key node method [21]. Besides the finite-element model of interest, another node is placed in the model, and the DOFs are 3, that is, d_i^{key} $(i = 1, 2, 3)$. These DOFs are called key DOFs and are regarded as equivalent to in-plane global strain E_{ij} as follows:

$$d_1^{key} = E_{11}, \, d_2^{key} = E_{22}, \, d_3^{key} = 2E_{12}$$

Eq. (5.9)

Therefore, by inputting the displacement value of the key node, the global strain can be applied to the finite-element model. This allows for a reasonable implementation of periodic unit cell simulation by relating the displacement gap of corresponding nodes on the periodic boundary to the displacement of the key node, which represents the global strain. For instance, the displacement gap along one axis between Edge (a) and Edge (c) is constrained by

$$u_1\big|_c - u_1\big|_a = Ld_1^{key} + \frac{L}{2}d_3^{key}$$

Eq. (5.10)

In this study, the equation function in ABAQUS/Standard is used to control the periodic boundary condition and the global strain defined by Eq. (5.10) and others.

5.1.6 Models for Matrix and Interface

This analysis considers matrix failure that is dependent on strain rate and stress triaxiality, with a constitutive relationship of elastoplastic deformation. The equivalent strain includes both elastic and plastic equivalent strain, and damage initiation is defined on the basis of the value of plastic equivalent strain. Figure 5.2 displays the definition of the damage initiation plastic strain, which depends on stress triaxiality and strain rate. The stress triaxiality is determined by dividing the mean stress by the equivalent stress. The equivalent plastic strain at damage initiation decreases with an increase in stress triaxiality, indicating dilatational stress damage. If the stress triaxiality exceeds 0.5, the equivalent plastic strain at damage initiation is assumed to be the same as that when the stress triaxiality is 0.5. For low stress triaxiality, failure occurs in the shear-dominant mode. For stress triaxiality lower than 0.167, the same value used when the stress triaxiality is 0.167 is employed. This definition can be implemented in ABAQUS, a commercial finite-element analysis software, without requiring any coding or user subroutine, making it easy to conduct strain-rate-dependent failure simulations. The temperature difference is accounted for by the reduced strain rate, which incorporates the strain rate difference. In ABAQUS, damage evolution can be defined on the basis of either energy-based failure or displacement-based failure. In this analysis, damage evolution is assumed to be completed almost immediately after

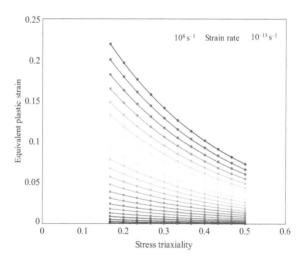

FIGURE 5.2 Damage initiation criteria defined by equivalent plastic strain as functions of reduced strain rate and stress triaxiality.

damage initiation, given the assumption of a relatively brittle resin matrix such as thermosetting epoxy resin.

To model interface strength, it is assumed that it is independent of strain rate (time) and temperature. A cohesive element is inserted into the fiber–matrix interface, and the cohesive traction separation behavior in mixed mode is shown in previous work [24]. The initiation of softening at the mixed mode is defined by

$$\left(\frac{\langle t_n \rangle}{Y_n}\right)^2 + \left(\frac{t_s}{Y_s}\right)^2 = 1. \qquad\qquad \text{Eq. (5.11)}$$

Here t_n is normal stress, t_s is shear stress, Y_n is pure tensile strength, Y_s is pure shear strength, and $\langle\rangle$ yields zero when t_n is a negative value. The toughness is defined as

$$\left(\frac{G_I}{G_{I\,C}}\right) + \left(\frac{G_{II}}{G_{II\,C}}\right) = 1 \qquad\qquad \text{Eq. (5.12)}$$

Here G_C is the critical energy release rate and G is the working energy release rate. Subscripts of I and II indicate Mode I and Mode II. In the present analysis, $\sqrt{2}Y_n = Y_s = 127\text{MPa}$ and $2G_{Ic} = G_{IIc} = 0.2\text{N}/\text{m}$ are assumed, on the basis of conventional works [24–27].

The quadratic criterion, as shown in Eq. (5.11), is often used to represent the failure envelope for interface strength. However, a quasi-parabolic criterion is a more precise representation as it accounts for the enhancement of shear strength under compressive stress on the interface. The failure envelope generated by the quasi-parabolic criterion is similar to that of the parabolic criterion in the first quadrant, as illustrated in Figure 2.9. Implementation of this criterion in ABAQUS involves setting friction in addition to the cohesive interface. It is important to note that the quasi-parabolic criterion provides a more accurate representation of the interface behavior compared to the quadratic criterion, which assumes a constant shear strength regardless of the compressive stress.

5.1.7 Results and Discussions

This section presents the numerical results obtained by employing the modeling techniques previously discussed. At a high strain rate, the interface strength is weaker than the matrix strength, leading to interfacial debonding as the initial failure mode, followed by additional interface failures. The interface cracks eventually connect through matrix failure, culminating in the ultimate failure. Conversely, at a low strain rate, the matrix failure takes place first, where the stress concentration is most pronounced. The damage then advances within the matrix, with the matrix crack occasionally being trapped at the interface. In such instances, the interface remains unbroken, while the matrix crack circumvents the fibers and expands into a more extensive crack. These failure patterns closely resemble the experimental results as illustrated in Figure 2.6.

Figure 5.3 presents the stress–strain curves, where nearly instantaneous failures are observed for both cases. Following the initial crack, which could be either matrix failure or interface failure, the strain does not increase substantially toward ultimate failure. Finally, Figure 5.4 displays the relationship between the global stress, corresponding to the specimen strength, and the reduced strain rate. At relatively low strain rates, the simulation exhibits a matrix failure dominant mode, and the specimen strength decreases as the strain rate decreases. In contrast, at high strain rates, the specimen strength remains constant and independent of the strain rate. This is attributed to the fact that failure is dominated by interfacial debonding, and the interface strength is unaffected by time or temperature.

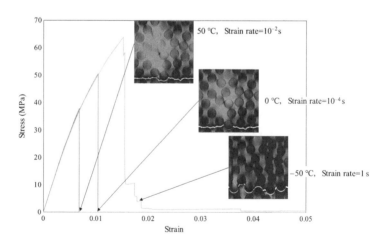

FIGURE 5.3   Stress–strain relationship for extremely high, low, and intermediate reduced strain rate.

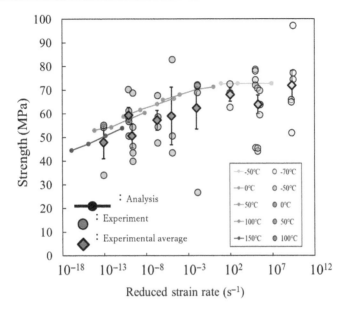

FIGURE 5.4   Simulated results of strength with reduced strain rates.

5.2 RESIDUAL STRENGTH OF UD COMPOSITE PREDICTED BY ENTROPY-BASED FAILURE CRITERION [28]

In this section, we incorporate an entropy-driven damage criterion into the matrix component of the periodic unit cell, exemplified in Figure 5.1. Previously, we had made a basic assumption that resin failure is governed by stress triaxiality and strain rate. As a result, the periodic unit cell analysis can model strain-rate and temperature dependencies. However, using the entropy-driven damage criterion (Eq. 3.24) allows us to numerically simulate various strength characteristics, including fatigue, creep, and residual strength after specific loads. Here, we model the residual strength after stress relaxation behavior of unidirectional CFRP in the transverse direction. It is worth mentioning that although many researchers have concentrated on residual strengths following specific loads, the residual strength after stress relaxation, which is analogous to a constant strain condition, has been overlooked and is assumed not to exhibit degradation. Nevertheless, envision a scenario where a specimen experiences constant strain just prior to failure. We can hypothesize that failure will eventually occur due to factors such as viscoelastic behavior.

The experimentally derived stress–strain relationships for neat resin polyimide specimens under uniaxial tensile loading at various temperatures and strain rates are presented in Figure 5.5. Predicted curves based on Eq. (3.24) are also plotted in this figure. The material constants utilized for this prediction can be found in Table 5.1. With this, the matrix modeling is finalized. The model is then integrated into a periodic unit cell consisting of 30 fibers and an encasing resin matrix, followed by the execution of a tensile simulation. This versatile model is capable of simulating a wide range of load cases, including creep, fatigue, constant strain, impact, and even combinations thereof.

In this study, we investigate the residual strength after a constant strain duration using both numerical simulations and experimental tests. The results are displayed in Figure 5.6. The specimens are unidirectional CFRP, and transverse tensile tests are conducted. Both standard tensile tests and

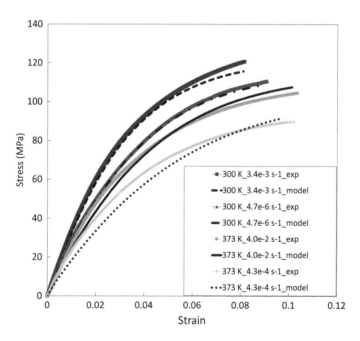

FIGURE 5.5 Experimental and numerical results of stress–strain curves for polyimide neat resin specimen under several strain rates and temperatures.

TABLE 5.1 Material Constants of Polyimide Resin Used in Figure 5.5, Figure 3.15, and Figure 3.16 [28]

MATERIAL PARAMETERS			
MAXWEL'S PARAMETERS		INITIAL YOUNG'S MODULUS	
n	E^n [MPa]	E_0 [MPa]	4,260
1~15	284	Nonlinear	
n	η^n [MPa·s]	σ_0 [MPa]	70
1	4.5×10^2	α	2
2	3.3×10^3	Damage parameters	
3	1.2×10^5	D^{cr}	0.1
4	1.9×10^6	γ^{cr} [kJ/m^{3}K]	15.5
5	1.8×10^7	Permanent strain	
6	1.4×10^8	η^{irr} [MPa·s]	1.0×10^{23}
7	8.5×10^8		
8	5.0×10^9		
9	3.0×10^{10}		
10	1.9×10^{11}		
11	1.4×10^{16}		
12	1.3×10^{19}		
13	2.1×10^{22}		
14	1.3×10^{26}		
15	2.5×10^{29}		

interrupted tensile tests are performed. In the interrupted tests, the specimens undergo a constant strain of 0.4% for two weeks. To expedite the viscoelastic behavior, the temperature is raised to 250°C. The tensile tests are carried out at ambient temperature. As demonstrated in Figure 5.6, stress relaxation occurs over time. After the constant strain duration, the tensile test resumes, and the ultimate strength is found to have decreased compared to the results of the standard tensile test. This observation implies that entropy is generated due to the viscoelastic behavior of the polyimide during the constant strain condition, which in turn leads to a reduction in residual strength. The outcomes of the numerical simulations are also illustrated in this figure. Similar to the experimental findings, the numerical simulations exhibit stress relaxation during the constant strain condition and a decrease in ultimate strength due to entropy generation.

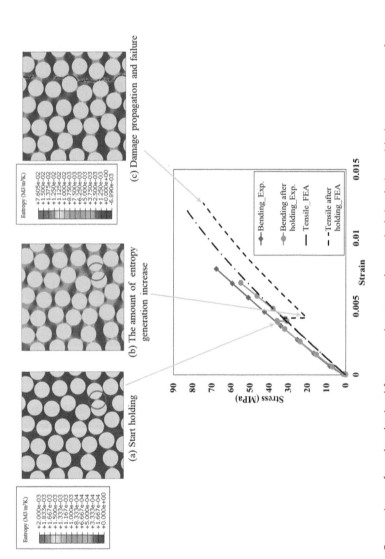

FIGURE 5.6 Comparison of results obtained from experiments and numerical simulation with damage process for normal bending test and interrupted tensile test with constant strain condition.

5.3 LONG-TERM DURABILITY FOR FIBER-DIRECTIONAL STRENGTH OF UD CFRP BASED ON SFF MODEL [29]

Ensuring long-term durability for composite materials requires a comprehensive understanding of the time and temperature dependence of fiber-directional strength in unidirectional carbon-fiber-reinforced polymers (UD-CFRPs). Prior to the failure of UD-CFRP in the fiber direction, even if numerous transverse cracks occur, they have little impact on the ultimate failure as long as the fiber-directional ply remains intact. The ultimate failure of composite materials is predominantly governed by fiber-directional failure, as unidirectional CFRP is often applied in load-bearing directions.

In Section 4.1 of this book, we introduce the model for predicting fiber-axial tensile strength of unidirectional composites, known as the SFF model. In this section, we apply the SFF model to predict the time- and temperature-dependent strengths of UD-CFRP. To use the SFF model, we need to input the strengths of the fiber, matrix, and interface. While the strengths of the fiber and interface are time- and temperature-independent, the matrix strength does exhibit time and temperature dependence. Thus, we must identify the time and temperature dependence of the matrix strength.

5.3.1 Matrix Strength

The elucidation of matrix strength's temperature dependence is of paramount importance and is executed in the following manner. We scrutinize the interdependence between time and temperature with respect to the relaxation modulus and tensile strength for the epoxy resin in question (Epikote 828, fabricated by Yuka Shell Epoxy, Inc.). An extensive exposition of the experimental methodology and results is presented in Ref. [29]. The unadulterated resin specimen is ultimately cured at 190°C for a duration of two hours.

To ascertain a master curve for the relaxation modulus, an initial bending creep test is performed at 40°C, followed by increments of 10°C from 60°C to 180°C, under relatively low-stress levels. This is premised on the assumption that the relaxation modulus is the inverse of creep compliance. Tensile strength assessments are conducted at testing velocities spanning from 0.05 mm/min to 50 mm/min and temperatures between 40°C and 160°C in 20°C intervals. Considering that the specimen is cured at 190°C and the test period is consistently brief, we postulate that the influence of physical aging on the resin's mechanical attributes is negligible within the purview of this investigation. Employing the subsequent equation, we ascertain the activation energy as:

$$\log\left[\alpha\left(T_{\text{test}}\right)\right] = \frac{\Delta H}{2.303 \cdot G}\left(\frac{1}{T_{\text{test}}} - \frac{1}{T_{\text{ref}}}\right)$$

Eq. (5.13)

where G is the gas constant 8.314×10^{-3} kJ/(Kmol), T_{test} is the test temperature (K), T_{ref} is the reference temperature (K), and ΔH is the activation energy. The reference temperature is 323 K (50°C). The activation energy for this material is 110 kJ/mol when the temperature is below 114°C and 600 kJ/mol when the temperature is above 114°C as reported in Ref. [29]. We approximate the relaxation-modulus master curve by employing

$$E(t) = \frac{1}{J_0\left\{1 + \left(\frac{t}{\tau_0}\right)^n\right\}}$$

Eq. (5.14)

at reference temperature. Here $J_0 = 1/3.2$ [1/GPa], $\tau_0 = 1,50,000$ min, and $n = 0.27$ are selected for fitting to the experimental data.

Subsequently, we delineate the CSR strength as a function of the reduced time-to-failure, adopting the same activation energy as in Figure 5.7. For this material, a relatively smooth CSR strength master curve is attained by utilizing

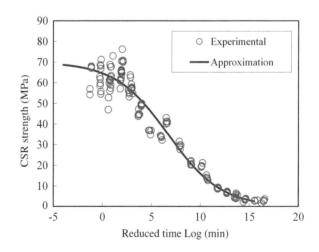

FIGURE 5.7 CSR strength master curve employing the same shift factor as modulus master curve (50°C reference temperature).

congruent shift factors. In the present study, the CSR strength as a function of reduced time is approximated via the empirical equation (5.15), incorporating the power law principle.

$$\sigma(t) = \frac{\sigma_0}{\left(1 + \left(\dfrac{t}{T_1}\right)^m\right)}$$ Eq. (5.15)

Here σ_0 is instantaneous strength, T_1 is equivalent to a specific degradation time, and m is the power law component. For the approximation in Figure 5.7, $\sigma_0 = 70$ MPa, $T_1 = 5{,}000{,}000$ min, and $m = 0.16$ are selected for fitting to the experimental data. Contemplating the shift factor and master curve in Eq. (5.15), the correlation between matrix strength and temperature is derived. It is imperative to note that the reference time is presumed to be 30 seconds. In other words, the matrix strength when the matrix fails in 30 seconds is a function of temperature.

5.3.2 Time and Temperature Dependence of UD CFRP Strengths Based on SFF Model

In our previous study [29], tensile strength under various temperature conditions was examined using two unidirectional composites. One composite, XN05, consisted of a relatively low modulus and weak carbon fiber, while the other, XN50, was made up of a relatively high modulus and strong carbon fiber (Nippon Graphite Fiber, Inc.). Both composites used the same surface treatments, matrix resin (Epikote 828), and curing conditions. Except for the fiber, the matrix and interface mechanical properties can be considered identical for these two composites. They are referred to as the XN05 composite and the XN50 composite.

The specimen and experiment procedure are described in detail in Ref. [29]. Figure 5.8 presents the results of the tests, with the vertical axis representing the composite strength, which is the load divided by the cross-sectional area of only fibers. To eliminate or minimize any effects of time-dependent characteristics, the tests were conducted in such a way that the specimen would fail within 30 seconds. The curves in the figures are SFF (Strength of Fiber Failure) predictions. The material constants used for this prediction are shown in Table 5.2. The interface strength is assumed to be shown in following equation.

$$\tau_i(T) = \tau_{const} + a \cdot \sigma_{i,thermal}(T)$$ Eq. (5.16)

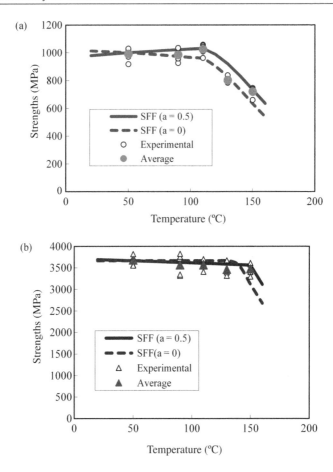

FIGURE 5.8 Experimental and SFF-estimated results of temperature-dependent tensile strength for (a) XN05 composite and (b) XN50 composite.

TABLE 5.2 Mechanical Constants Used for the SFF [29]

PROPERTY		UNIT	XN05 COMPOSITE	XN50 COMPOSITE
L_0		mm	25	25
σ_0		MPa	940	3,400
m			7.5	7.5
r		mm	0.005	0.005
V_f			0.55	0.55
τ_{const}	$a = 0.5$	MPa	11	11
	$a = 0$		15.6	15.6

Figure 5.9 presents the experimental results and the Strength of Fiber Failure (SFF) estimated results for tensile strengths of both XN05 and XN50 composites. It represents the relationship between strength and reduced time-to-failure. For both Carbon-Fiber-Reinforced Polymers (CFRPs), the

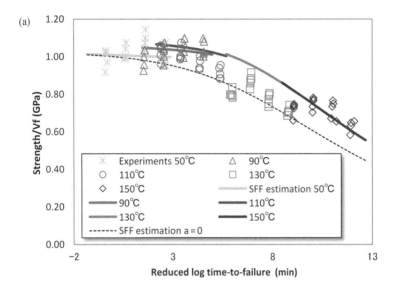

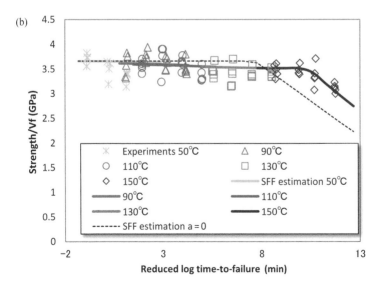

FIGURE 5.9 Comparisons of SFF estimations and experimental results of (a) XN05 composite and (b) XN50 composite.

SFF predictions align well with the experimental results. This demonstrates that our proposed method can reasonably estimate the time and temperature dependence of CFRP strengths. However, it's important to note that this estimation is only applicable for a constant strain test. If we use an entropy-based matrix strength in conjunction with the SFF model, we can estimate a wide range of CFRP strengths, including fatigue strength, creep strength, and impact strength, as well as combinations of these. This suggests that our method can provide a comprehensive analysis of CFRP performance under various conditions.

REFERENCES

[1] J. Koyanagi, S. Yoneyama, A. Nemoto, J.D.D. Melo, Time and temperature dependence of carbon/epoxy interface strength, Composites Science and Technology 70(9) (2010) 1395–1400.

[2] J. Koyanagi, H. Hatta, F. Ogawa, H. Kawada, Time-dependent reduction of tensile strength caused by interfacial degradation under constant strain duration in UD-CFRP, Journal of Composite Materials 41(25) (2016) 3007–3026.

[3] J. Koyanagi, M. Nakada, Y. Miyano, Prediction of long-term durability of unidirectional CFRP, Journal of Reinforced Plastics and Composites 30(15) (2011) 1305–1313.

[4] J. Koyanagi, M. Nakada, Y. Miyano, Tensile strength at elevated temperature and its applicability as an accelerated testing methodology for unidirectional composites, Mechanics of Time-Dependent Materials 16(1) (2011) 19–30.

[5] J. Koyanagi, S. Ogihara, Temperature dependence of glass fiber/epoxy interface normal strength examined by a cruciform specimen method, Composites Part B: Engineering 42(6) (2011) 1492–1496.

[6] J. Koyanagi, S. Yoneyama, K. Eri, P.D. Shah, Time dependency of carbon/epoxy interface strength, Composite Structures 92(1) (2010) 150–154.

[7] A. Straub, M. Slivka, P. Schwartz, A study of the effect of time and temperature on the fiber/matrix interface using the microbond test, Composites Science and Technology 57 (1997) 991–994.

[8] J. Koyanagi, A. Yoshimura, H. Kawada, Y. Aoki, A numerical simulation of time-dependent interface failure under shear and compressive loads in single-fiber composites, Applied Composite Materials 17(1) (2010) 31–41.

[9] M. Herráez, C. González, C.S. Lopes, R.G. de Villoria, J. Llorca, T. Varela, J. Sánchez, Computational micromechanics evaluation of the effect of fibre shape on the transverse strength of unidirectional composites: An approach to virtual materials design, Composites Part A: Applied Science and Manufacturing 91 (2016) 484–492.

[10] M. Herráez, D. Mora, F. Naya, C.S. Lopes, C. González, J. Llorca, Transverse cracking of cross-ply laminates: A computational micromechanics perspective, Composites Science and Technology 110 (2015) 196–204.

[11] J. Koyanagi, Y. Sato, T. Sasayama, T. Okabe, S. Yoneyama, Numerical simulation of strain-rate dependent transition of transverse tensile failure mode in fiber-reinforced composites, Composites Part A: Applied Science and Manufacturing 56 (2014) 136–142.

[12] C.S. Lopes, C. González, O. Falcó, F. Naya, J. Llorca, B. Tijs, Multiscale virtual testing: the roadmap to efficient design of composites for damage resistance and tolerance, CEAS Aeronautical Journal 7(4) (2016) 607–619.

[13] F. Naya, C. González, C.S. Lopes, S. Van der Veen, F. Pons, Computational micromechanics of the transverse and shear behavior of unidirectional fiber reinforced polymers including environmental effects, Composites Part A: Applied Science and Manufacturing 92 (2017) 146–157.

[14] F. Naya, M. Herráez, C.S. Lopes, C. González, S. Van der Veen, F. Pons, Computational micromechanics of fiber kinking in unidirectional FRP under different environmental conditions, Composites Science and Technology 144 (2017) 26–35.

[15] S.A. Elnekhaily, R. Talreja, Damage initiation in unidirectional fiber composites with different degrees of nonuniform fiber distribution, Composites Science and Technology 155 (2018) 22–32.

[16] H. Ghayoor, S.V. Hoa, C.C. Marsden, A micromechanical study of stress concentrations in composites, Composites Part B: Engineering 132 (2018) 115–124.

[17] L. Riaño, L. Belec, Y. Joliff, Validation of a representative volume element for unidirectional fiber-reinforced composites: Case of a monotonic traction in its cross section, Composite Structures 154 (2016) 11–16.

[18] S.H.R. Sanei, R.S. Fertig, Uncorrelated volume element for stochastic modeling of microstructures based on local fiber volume fraction variation, Composites Science and Technology 117 (2015) 191–198.

[19] H. Yuanchen, J. Kyo Kook, H. Sung Kyu, Effects of fiber arrangement on mechanical behavior of unidirectional composites, Journal of Composite Materials 42(18) (2008) 1851–1871.

[20] S. Li, N. Warrior, Z. Zou, F. Almaskari, A unit cell for FE analysis of materials with the microstructure of a staggered pattern, Composites Part A: Applied Science and Manufacturing 42(7) (2011) 801–811.

[21] K. Terada, N. Hirayama, K. Yamamoto, S. Matsubara, Numerical plate testing for linear multi-scale analyses of composite plates, International Journal for Numerical Methods in Engineering 105 (2016) 111–137.

[22] K. Yoshida, M. Nakagami, Numerical analysis of bending and transverse shear properties of plain weave fabric composite laminates considering the intra-lamina inhomogeneity, Journal of the Japan Society for Composite Materials 42(1) (2016) 2–12.

[23] A. Schmitz, P. Horst, A finite element unit-cell method for homogenised mechanical properties of heterogeneous plates, Composites Part A: Applied Science and Manufacturing 61 (2014) 23–32.

[24] P.D.S. Jun Koyanagi, Souta Kimura, Sung K Ha, Hiroyuki Kawada, Mixed-mode interfacial debonding simulation in single fiber composite under transverse load, Journal of Solid Mechanics and Materials Engineering 3 (2009) 796–806.

[25] J. Koyanagi, H. Nakatani, S. Ogihara, Comparison of glass–epoxy interface strengths examined by cruciform specimen and single-fiber pull-out tests under combined stress state, Composites Part A: Applied Science and Manufacturing 43(11) (2012) 1819–1827.

[26] J. Koyanagi, S. Ogihara, H. Nakatani, T. Okabe, S. Yoneyama, Mechanical properties of fiber/matrix interface in polymer matrix composites, Advanced Composite Materials 23(5–6) (2014) 551–570.

[27] M. Sato, J. Koyanagi, X. Lu, Y. Kubota, Y. Ishida, T.E. Tay, Temperature dependence of interfacial strength of carbon-fiber-reinforced temperature-resistant polymer composites, Composite Structures (2018) 202 (2018), 283–289.

[28] M. Sato, K. Hasegawa, J. Koyanagi, R. Higuchi, Y. Ishida, Residual strength prediction for unidirectional CFRP using a nonlinear viscoelastic constitutive equation considering entropy damage, Composites Part A: Applied Science and Manufacturing 141 (2021).

[29] J. Koyanagi, M. Nakada, Y. Miyano, Tensile strength at elevated temperature and its applicability as an accelerated testing methodology for unidirectional composites, Mechanics of Time-Dependent Materials 16(1) (2012) 19–30.

Durability Predicted by Mesoscale Simulations

6

6.1 DEVELOPMENT OF FINITE ELEMENT MODEL FOR SIMULATING ENTROPY-BASED STRENGTH-DEGRADATION FOR CFRP SUBJECTED TO VARIOUS LOADINGS

Given the escalating ubiquity of carbon-fiber-reinforced plastics (CFRPs), it becomes paramount to emphasize the material's durability, in alignment with the global Sustainable Development Goals (SDGs). As such, the long-term dependability of CFRPs often necessitates assessment through repeated loading tests [1–13]. Various scholars have delved into the fatigue properties of CFRP, elucidating topics ranging from fatigue-induced failure under consistent cyclic loading to the onset and propagation of transverse cracks. While experimental analyses remain a cornerstone for fatigue studies in CFRPs, there exists a burgeoning need for validated numerical methodologies that would predict the actual long-term durability of CFRPs in service accurately. We are of the belief that our understanding of residual strength post-fatigue under generalized arbitrary conditions should facilitate accurate predictions via modeling, thereby bypassing the need for exhaustive experimentation. Such a model would expedite the design process of CFRP structures and components. Furthermore, numerical fatigue simulation techniques that can accommodate varying stress ratios and frequencies are deemed advantageous.

DOI: 10.1201/9781003371137-6

CFRP failure is typically an accumulation of microfailures in the fiber, matrix, and fiber/matrix interface. Interestingly, the strength of the fiber and interface are independent of time, temperature, and stress-history, making the matrix properties the primary determinant of the macro-properties of CFRPs in terms of their time, temperature, and stress-history dependence [14–16]. Thus, when considering CFRP durability, it is crucial to investigate the matrix properties under the aforementioned variables. Consequently, the matrix's nonlinear constitutive relationship, inclusive of damage and failure, becomes the primary determinant of CFRP's long-term reliability. Several past studies have defined matrix strength as a function of stress triaxiality and strain rate to simulate CFRP's strain-rate dependence [17, 18]. Although applicable to monotonic loading conditions, these definitions fail under cyclic or other complex loading conditions. On the other hand, an elasto-viscoplastic constitutive relationship can accommodate varying loading conditions based on damage involving the plastic strain rate [19–22]. However, one potential limitation of these constitutive relationships is that damage does not progress unless a new strain increment is imposed. Consequently, time-dependent behaviors including damage progression, primarily due to viscoelasticity, may not be perfectly simulated, raising questions about their suitability for discussing long-term durability of CFRPs.

In recent years, entropy-based failure criteria have found application in matrix resins and some CFRPs. Herein, material failure is determined by critical entropy generation values [9, 23–28]. Intact materials have a relatively low entropy due to a well-ordered molecular alignment. Upon application of a thermomechanical load, this molecular alignment partially collapses, leading to a simultaneous increase in entropy. This collapse equates to material inelastic deformation, suggesting imminent failure. In cyclic loadings, molecular alignment collapses progressively, leading to a gradual entropy increase and eventual material failure. Mechanically, entropy generation is ascertained by dividing dissipated energy by absolute temperature per unit volume. Assuming a constant temperature, entropy generation is proportional to dissipated energy, which can be calculated from the experimentally obtained stress–strain relationship during mechanical testing. In nonexperimental scenarios, the dissipated energy can be calculated if the inelastic constitutive equation is known. To discuss CFRP failure at the macroscopic level, a microscale model distinguishing between fiber and matrix must be augmented to a macroscale model, treating the ply as a homogeneous material. In other words, the constitutive relationship, including the damage and failure of a ply in CFRPs, is dominated on the basis of the matrix properties. The current study does not focus on the correlation method between these scales, as it has already been established. Instead, it outlines the development of a macroscale model while forgoing the scale-up procedure.

The present model leverages the three-dimensional Hashin criterion, defining the damage onset stress as material strength, with degradation occurring in sync with entropy generation [29]. Therefore, the fracture energy decreases alongside entropy generation. In this model, an orthotropic viscoelastic constitutive relationship is established, encompassing damage and strength reduction, with the dissipated energy calculated from the work expended during viscous deformations. This model affords independent definition of strength reduction in each direction (fiber-directional tensile, fiber-directional compression, transverse tensile, transverse compression, fiber-axial shear, and fiber-transverse shear), based on entropy generation. Since the entropy generation at the failure of the matrix resin remains constant and independent of the loading direction, the entropy generation is ascertained by summing the dissipated energy across all directions.

The model's parameters are partially determined on the basis of microscale analysis. However, we anticipate refining these parameters in future studies, and, currently, we employ provisional specific values to verify the model's behavior. This model elucidates strength and fracture energy reduction based on the stress–strain history, simulating strength reduction and delayed failure under conditions of constant stresses, constant strains, cyclic loadings, and fatigue levels arising from varying stress levels and frequencies. In utilizing this model, researchers can simulate the initiation and evolution of transverse cracking in CFRP laminates and predict fatigue failure in random loadings, among other possibilities. It is our expectation that such a comprehensive approach will prove valuable in enhancing our understanding and utilization of CFRP in various fields.

6.2 NUMERICAL PROCEDURES [29]

6.2.1 Orthotropic Viscoelastic Relaxation Modulus

First the following equation is employed to express the relaxation modulus of the resin matrix [14, 30]:

$$E_{\mathrm{m}}(t) = \frac{E_{\mathrm{m},0}}{1 + \left(\dfrac{t}{T_{\mathrm{c}}}\right)^{n}} \qquad \text{Eq. (6.1)}$$

Here $E_{m,0}$ is the initial elastic modulus, t is the time, T_c is the specific time, and n is the power law component. $E_{m,0} = 3,000$ MPa, $T_c = 5,000,000$, and n = 0.3 were employed in this study, which are intended for a general epoxy resin. Based on the rule of mixture, the relaxation moduli in six directions (E_{11}, E_{22}, E_{33}, G_{12}, G_{13}, and G_{23}) for CFRP materials were determined.

$$E_{ii}(t) = \begin{cases} V_f E_{f,ii} + (1 - V_f) E_m(t) & (i = 1) \\ \dfrac{1}{\dfrac{V_f}{E_{f,ii}} + \dfrac{1 - V_f}{E_m(t)}} & (i = 2,3) \end{cases} \qquad \text{Eq. (6.2)}$$

$$G_{ij}(t) = \frac{1}{\dfrac{V_f}{G_{f,ij}} + \dfrac{1 - V_f}{G_m(t)}} \qquad \text{Eq. (6.3)}$$

In the present examination, these are contemplated as the calculated relaxation moduli. We adopt the material attributes of the matrix resin and carbon fiber as reported in References [14, 30]. Simultaneously, we employ the numerical representation of five parallelly aligned Maxwell models to characterize the viscoelastic behavior under study. This numerical model constitutes five parallel-connected Maxwell models, with each Maxwell model comprising a spring tensor C_d^k and a dashpot tensor η^k. Stress can be determined by the summation of all elastic stresses $\tilde{A}^k$. In the subsequent discourse, superscript k will denote the value corresponding to the kth Maxwell model. Stress relaxation is perceived as an independent occurrence in each line. For orthotropic material properties, we must independently define six directional stress–strain relationships. This necessitates the involvement of 30 elasticity constants and 30 viscosity constants in total. For this study, we have opted to use the material parameters as illustrated in Table 6.1. This choice was driven by the comparison between the analytical curves derived using Eqs (6.2) and (6.3) and the numerical results gleaned from the model assumed in this study, particularly in relation to the relaxation modulus. These values essentially aim to encapsulate a generic composite material composed of carbon fibers and epoxy resin [18]. The values of η_{ij}^1 are extremely high, which implies an almost elastic behavior. In terms of Poisson's ratio, $v_{12} = v_{13} = 0.34$ and $v_{23} = 0.4$. Though the alignment between analytical and numerical curves could be enhanced by increasing the number of Maxwell models, this falls outside the purview of the present study

TABLE 6.1 Material Properties of Springs and Dashpots for Viscoelastic Model [29]

UNIT (MPA)	E_{11}^k	E_{22}^k, E_{33}^k	G_{12}^k, G_{13}^k	G_{23}^k
$k = 1$	1,28,000	4,290	1,810	1,610
$k = 2$	80	267	133	101
$k = 3$	80	267	133	101
$k = 4$	80	267	133	101
$k = 5$	80	267	133	101
MPa·s	η_{11}^k	η_{22}^k, η_{33}^k	η_{12}^k, η_{13}^k	η_{23}^k
$k = 1$	1E+30	1E+30	1E+30	1E+30
$k = 2$	3.50E+06	1.17E+07	5.83E+06	4.45E+06
$k = 3$	3.00E+06	1.00E+07	5.00E+06	3.81E+06
$k = 4$	3.00E+05	1.00E+06	5.00E+05	3.81E+05
$k = 5$	6.00E+03	2.01E+04	9.99E+03	7.63E+03

and therefore will not be pursued. Instead, our focus is on determining the orthotropic viscoelastic relaxation modulus.

6.2.2 Implementation of Orthotropic Viscoelastic Constitutive Relation and Entropy Calculation in FEM

Figure 6.1 presents the flowchart delineating the stress-updating process from the (n−1)th step to the nth step. In this investigation, we employ the user subroutine UMAT tailored for the widely used finite element analysis (FEA) software, ABAQUS. For every temporal increment, we account for stress relaxation, concurrently quantifying the dissipated energy, and then proceed to implement the reductions in strength and fracture energy via the stiffness tensor. First, the total strain increment $\Delta\mu_t$ calculated in each step is decomposed into elastic component $\Delta\mu_e^k$ and viscoelastic component $\Delta\mu_v^k$ per each Maxwell model.

$$\Delta\varepsilon_{t,n} = \Delta\varepsilon_{e,n}^k + \Delta\varepsilon_{v,n}^k \qquad\qquad \text{Eq. (6.4)}$$

Analytical curve
Fitted curve

The incremental form of the elastic constitutive law is expressed as

$$\Delta\sigma_n^k = C_{d,n}^k \Delta\varepsilon_{e,n}^k$$ Eq. (6.5)

Here, the damaged stiffness tensor of k-th spring $C_{d,n}^k$ is defined as

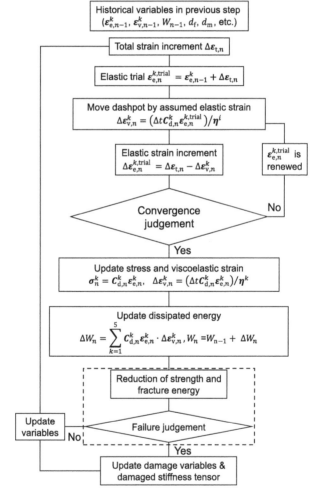

FIGURE 6.1   Overall flowchart for updating stress, entropy, strength, damage variables, and stiffness tensor.

$$
\boldsymbol{C}_{d,n}^{k} =
\begin{bmatrix}
(1-d_{f,n})C_{11}^{k} & (1-d_{f,n})(1-d_{m,n})C_{12}^{k} & (1-d_{f,n})(1-d_{m,n})C_{13}^{k} & 0 & 0 & 0 \\
 & (1-d_{m,n})C_{22}^{k} & (1-d_{m,n})C_{23}^{k} & 0 & 0 & 0 \\
 & & (1-d_{m,n})C_{33}^{k} & 0 & 0 & 0 \\
 & & & (1-d_{f,n})(1-d_{m,n})C_{44}^{k} & 0 & 0 \\
 & Sym. & & & (1-d_{f,n})(1-d_{m,n})C_{55}^{k} & 0 \\
 & & & & & (1-d_{m,n})C_{66}^{k}
\end{bmatrix},
$$

$$\text{Eq. (6.6)}$$

where $d_{f,n}$ and $d_{m,n}$ are fiber and matrix damage variables defined in the next section, and C_{ij}^{k} is the component of stiffness tensor of intact material which is calculated using material properties in Table 6.1. The increment of viscous component in Eq. (6.4) is expressed as

$$
\Delta\varepsilon_{v,n}^{k} = \frac{C_{d,n}^{k}\varepsilon_{e,n}^{k}}{\eta^{k}}\Delta t
\qquad\text{Eq. (6.7)}
$$

As a trial step, the total strain increment is first assumed to be only composed of elastic strain (i.e., $\Delta\varepsilon_{v,n}^{k} = 0$). The first trial stress is then

$$
\sigma_{n}^{k,\text{trial}} = C_{d,n}^{k}\varepsilon_{e,n}^{k,\text{trial}} = C_{d,n}^{k}\left(\varepsilon_{e,n-1}^{k} + \Delta\varepsilon_{t,n}\right)
\qquad\text{Eq. (6.8)}
$$

The stress delineated facilitates movement within the dashpot over the time increment, thereby inducing $\Delta\varepsilon_{v}^{k}$, which consequently causes the elastic stress to evolve to $C_{d,n}^{k}\left(\varepsilon_{e,n-1}^{k} + \Delta\varepsilon_{t,n}\right)$, referred to as the second trial stress. Following this, we proceed to examine the movement of the dashpot under this stress. This iterative process continues until the ratio between the newly computed stress and the preceding stress falls within a ±0.01% convergence range.

Following this, an update of the elastic stress takes place, and, concurrently, the components of Eq. (6.4) are ascertained.

Subsequent to this, utilizing Eq. (6.8), the incremental dissipated energy ΔW_n and accumulated dissipated energy W_n undergo an updating process as follows.

$$\Delta W_n = \sum_{k=1}^{5} C_{d,n}^k \varepsilon_{e,n}^k \cdot \Delta \varepsilon_{v,n}^k,$$ Eq. (6.9)

$$W_n = W_{n-1} + \Delta W_n,$$ Eq. (6.10)

An essential observation to be made is that the dissipated energies across all directions are aggregated for the assessment of entropy, aligning with our previous study [24]. For instance, W is equivalent to the area of the hysteresis loop of stress–strain history in the case of cyclic loading. The accumulated dissipated energy, upon being divided by the absolute temperature, indicates entropy generation ($s = W/T$). This generation of entropy is linked with strength degradation and fracture energy degradation, as illustrated in Table 6.2. Damage onset (initiation) at the n_0-th time step is taken into consideration here. In Table 6.2, X_T, X_C, Y_T, Y_C, S_{12} (= S_{13}), and S_{23} represent the strengths in the axial tensile, axial compressive, transverse tensile, transverse compressive, axial (in-plane) shear, and transverse shear directions, respectively. G_{ft}, G_{fc}, G_{mt}, and G_{mt} are the fracture toughness in axial tensile, axial compressive, transverse tensile, and transverse compressive. Subscript 0 represents their initial values. α_{AT}, α_{AC}, and α_O are arbitrary constants which may be a function of entropy s; s is entropy generation which can be determined dividing W by temperature. They are important and govern the correlation between resin entropy damage and the degradation of ply properties. A more comprehensive study, outside the scope of the current study, is required to discuss α (α_{AT}, α_{AC} and α_O). For instance, micromechanical modeling could be utilized to establish a correlation between microscale damage and macroscale damage, as presented by Sawamura et al. [18]. However, the required procedure is not trivial. Therefore, the discussion regarding this procedure is excluded herein, and a constant value of $\alpha_{AT} = \alpha_{AC} = \alpha_O = 300{,}000$ Kmm^3/J is applied. An empirical correlation between resin entropy damage and ply-property degradation will be the subject of investigation in a subsequent study.

Moreover, we incorporate Hashin's failure criteria, as utilized in the authors' prior study [18]. Four damage onset criteria in the fiber-directional tensile mode (e_{ft}), fiber-directional compressive mode (e_{fc}), transverse directional tensile mode (e_{mt}), and transverse directional compressive mode (e_{mc})

TABLE 6.2 Reductions of Strength and Fracture Energy and Failure Criteria

Strength reduction	$X_{T,n_0} = \left(1 - \alpha_{AT} s_{n_0}\right) X_{T,0}$, $X_{C,n_0} = \left(1 - \alpha_{AC} s_{n0}\right) X_{C,0}$, $Y_{T,n_0} = \left(1 - \alpha_O s_{n0}\right) Y_{T,0}$, $Y_{C,n_0} = \left(1 - \alpha_O s_{n0}\right) Y_{C,0}$, $S_{12,n_0} = \left(1 - \alpha_O s_{n0}\right) S_{12,0}$, $S_{13,n_0} = \left(1 - \alpha_O s_{n0}\right) S_{13,0}$, $S_{23,n_0} = \left(1 - \alpha_O s_{n0}\right) S_{23,0}$
Fracture energy reduction	$G_{ft,n_0} = \left(1 - \alpha_{AT} s_{n0}\right)^2 G_{ft,0}$, $G_{fc,n_0} = \left(1 - \alpha_{AC} s_{n0}\right)^2 G_{fc,0}$, $G_{mt,n_0} = \left(1 - \alpha_O s_{n0}\right)^2 G_{mt,0}$, $G_{mc,n_0} = \left(1 - \alpha_O s_{n0}\right)^2 G_{mc,0}$
Failure criteria	$e_{ft} = \left(\dfrac{\sigma_{11}}{X_{T,n_0}}\right)^2 + \left(\dfrac{\tau_{12}}{S_{12,n_0}}\right)^2 + \left(\dfrac{\tau_{13}}{S_{13,n_0}}\right)^2$ $e_{fc} = \left(\dfrac{\sigma_{11}}{X_{C,n_0}}\right)^2$ $e_{mt} = \left(\dfrac{\sigma_{22} + \sigma_{33}}{Y_{T,n_0}}\right)^2 + \dfrac{\tau_{23}^2 - \sigma_{22}\sigma_{33}}{\left(S_{23,n_0}\right)^2} + \dfrac{\tau_{12}^2}{\left(S_{12,n_0}\right)^2} + \dfrac{\tau_{13}^2}{\left(S_{13,n_0}\right)^2}$ $e_{mc} = \left[\left(\dfrac{Y_{C,n_0}}{2S_{23,n_0}}\right)^2 - 1\right]\dfrac{\sigma_{22} + \sigma_{33}}{Y_{C,n_0}} + \left(\dfrac{\sigma_{22} + \sigma_{33}}{2S_{23,n_0}}\right)^2$ $+ \dfrac{\tau_{23}^2 - \sigma_{22}\sigma_{33}}{\left(S_{23,n_0}\right)^2} + \dfrac{\tau_{12}^2}{\left(S_{12,n_0}\right)^2} + \dfrac{\tau_{13}^2}{\left(S_{13,n_0}\right)^2}$

are calculated using deteriorated strengths. If any criterion is satisfied (i.e., $e \geq 1$), it signifies the onset of damage, and the damage evolution algorithm is applied.

A detailed algorithm for damage evolution and a schematic illustration of the proposed constitutive model are shown in Figure 6.4 and Figure 6.5, respectively. Let's discuss the case of fiber tensile failure as an example. The damage onset and evolution algorithms mirror those of the conventional energy-regularized damage model, except for the strength and fracture energy degradation. As demonstrated in Figure 6.3, the strength and fracture energy deteriorate concurrently based on entropy generation. While the stress–strain relationship attempts to maintain its initial shape resemblance, the shape may change when entropy generation occurs during damage evolution, thus reducing the remaining fracture energy. Damage variables are computed on the basis of the change in fracture energy, as determined by the changes in equivalent stress and relative displacement, as depicted in Figure 6.2.

A crucial point about this damage evolution algorithm is that the damage evolves over time, even when displacement (or strain) ceases. Without implementing this algorithm, which leads to fracture energy degradation during the damage evolution period, there may be scenarios where the element stiffness

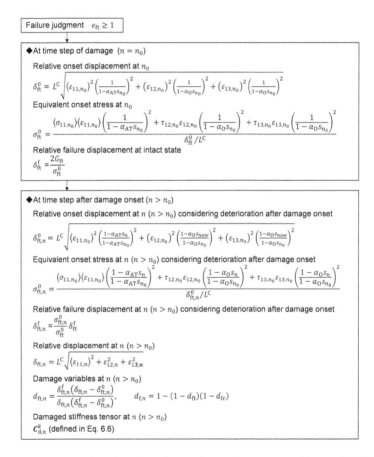

FIGURE 6.2 Detailed Flowchart for updating damage variables and stiffness tensor considering material degradation after damage onset in the case of fiber tensile failure.

doesn't drop off even if cyclic loading is applied. In such cases, simulating transverse crack propagation for CFRP laminates under cyclic loading conditions might not be possible because if the element stiffness is preserved, neighboring elements might not fail. The algorithm of time-dependent degradation must be applied not only to the damage onset criterion but also to the damage evolution criterion. Implementing this algorithm can enable numerical simulation of an increase in transverse cracks in laminated CFRPs under displacement-controlled fatigue. Table 6.3 shows the variables used in this study, some of which are identical to those used in the authors' previous study [31].

6.3 EXPERIMENTAL [32]

6.3.1 Experimental Conditions

In this study, we used carbon fiber/epoxy unidirectional (UD) tape prepreg (Torayca, T700SC/2592, 0.14 mm/ply) as the material. The prepregs are cured in an autoclave at a temperature of 130°C and at a pressure of 0.2 MPa. The stacking configuration is $[0°/90°_3]s$, with the post-cured thickness of the laminate being approximately 1.15 mm. The specimen's dimensions are shown in Figure 6.4. Laminates are cut to the required dimensions using a composite material cutting machine (AC-300CF, Maruto Testing Machine). GFRP tabs are glued to the ends of the specimen using adhesive glue before testing. Since CFRP laminates are opaque, crack observation is carried out using X-ray radiography.

For this purpose, an X-ray machine, M-100S, SOFTEX, is used with an applied voltage and current of 14 kVp and 1.5 mA, respectively, with an exposure time of 3 min. This damage observation method is a standard technique for detecting transverse cracks and delamination in CFRP laminates.

In order to establish the stress ratio used for fatigue testing in this study, six specimens from the same manufacturing batch were subjected to both monotonic and cyclic loading. These experiments were performed at a cross-head displacement speed of 1 mm/min using a Tensilon RTF-1350 A&D tensile testing machine. From these tests, the average maximum tensile strength of the laminate was found to be 647 MPa, with transverse cracks initiating at about the stress level of 250 MPa.

The fatigue test of the CFRP was carried out at room temperature. The relationships between transverse cracks and various load conditions, including cyclic load number and frequency, were examined. The fatigue loading

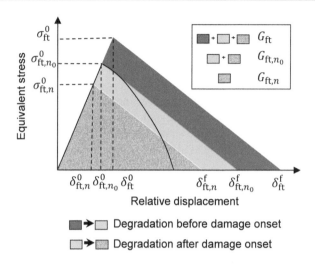

FIGURE 6.3 Schematic overview of constitutive relationship under axial loading considering strength and energy degradation due to entropy generation (dissipated energy divided by absolute temperature).

machine used in this study (Shimadzu, EHF-LV020K1A) is employed. The tension-tension sinusoidal fatigue load is applied during the tests, in which the stress ratio R (defined as $R = \sigma_{min}/\sigma_{max}$) is fixed at 0.1, where $\sigma_{min} = 20$ MPa and $\sigma_{max} = 200$ MPa (30% of the tensile strength). The transverse crack density in this study is calculated by dividing the number of transverse cracks by a specific area of 60 mm in length located at the center of the specimen.

6.3.2 Experimental Results

Figure 6.5 illustrates the typical distribution of transverse cracks under various cyclic loads, all with a fixed load frequency of 5 Hz. The results show that no transverse cracks appear when the cyclic load number is less than 10^3. However, as the cyclic load number increases to 10^4, two distinct transverse cracks are observed. The fatigue test is terminated once the cyclic load number reaches 10^5, at which point the number of transverse cracks has increased to 31.

In addition, Figure 6.6 displays the behavior of transverse cracks after 10^5 cyclic loads at various load frequencies. Contrary to the effect of the

TABLE 6.3 Material Properties Assumed in This Study [29]

MATERIAL PROPERTIES	SYMBOL	VALUE	UNITS
Initial axial tensile strength	$X_{T,0}$	3,930	MPa
Initial axial compressive strength	$X_{C,0}$	2,775	MPa
Initial transverse tensile strength	$Y_{T,0}$	150	MPa
Initial transverse compressive strength	$Y_{C,0}$	270	MPa
Initial axial shear strength	$S_{12,0}, S_{13,0}$	117	MPa
Initial transverse shear strength	$S_{23,0}$	117	MPa
Initial fiber-directional tensile fracture energy	$G_{ft,0}$	112.7	N/mm
Initial fiber-directional compressive fracture energy	$G_{fc,0}$	25.9	N/mm
Initial transverse tensile fracture energy	$G_{mt,0}$	0.5	N/mm
Initial transverse compressive fracture energy	$G_{mc,0}$	0.5	N/mm
Degradation coefficient (unique value for all component in this study)	α $(\alpha_{AT}, \alpha_{AC}, \alpha_0)$	300,000	Kmm³/J

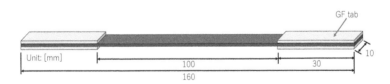

FIGURE 6.4 Specimen used for fatigue test.

cyclic load number, the final number of transverse cracks decreases from 37 to 24 as the load frequency increases from 1 Hz to 10 Hz.

These fatigue test results lead to the conclusion that both the cyclic load number and the load frequency significantly influence the formation of transverse cracks in CFRP. These factors will be further investigated in the following section, using a novel numerical method based on an entropy-based failure criterion.

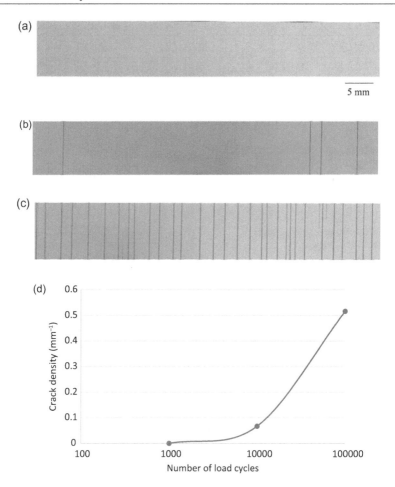

FIGURE 6.5   Transverse crack distribution under various cyclic load numbers at a maximum stress of 30% of the tensile strength: (a) 103 cycles: 0 transverse crack; (b) 104 cycles: 4 transverse cracks; (c) 105 cycles: 31 transverse cracks; and (d) effect of load number on crack density.

These fatigue test results lead to the conclusion that both the cyclic load number and the load frequency significantly influence the formation of transverse cracks in CFRP. These factors will be further investigated in the following section, using a novel numerical method based on an entropy-based failure criterion.

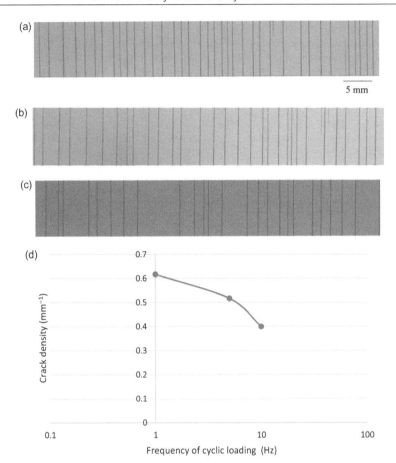

FIGURE 6.6 Transverse crack distribution under various load frequency (105 cycles): (a) Load frequency 1 Hz: 37 transverse cracks; (b) load frequency 5 Hz: 31 transverse cracks; (c) load frequency 10 Hz: 24 transverse cracks; and (d) effect of load frequency on crack density.

6.4 COMPARISON WITH NUMERICAL RESULTS

The proposed methodology is employed to simulate the transverse cracking behavior in CFRP structures. As depicted in Figure 6.7(a), the CFRP structure measures 100 mm × 10 mm × 6 mm, with symmetrical boundary conditions applied on surfaces ABCD, CDHG, and AEHD. Within the 90° layer, the transverse tensile strength is presumed to adhere to the cumulative distribution function for the Weibull distribution, given by the equation $\sigma = \sigma_0 \left(\ln \dfrac{1}{R-1} \right)^{1/m}$, where $m = 20$ and $\sigma_0 = 90$ MPa. A finite element model, consisting of 7,777 nodes and 6,000 C3D8 elements, is presented in Figure 6.7(b). The color in this figure denotes the material property, and the size of each element is 1 mm × 1 mm × 1 mm. In this example, the transverse crack density is calculated by dividing the number of transverse cracks by the length of the structure (100 mm). To ensure numerical stability, the stress boundary condition is substituted with strain boundary conditions. This means

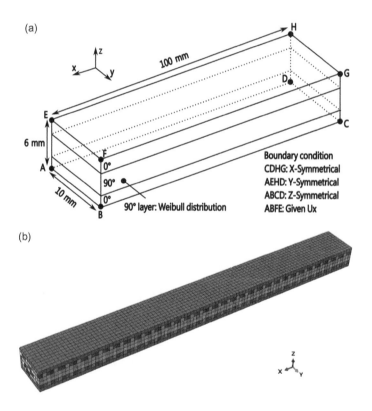

FIGURE 6.7 Cross-ply CFRP structures subjected to cyclic loading: (a) Geometric model under cyclic loading: symmetric boundary conditions are applied on faces ABCD, CDHG, and AEHD, respectively and (b) finite element model (color denotes the material property).

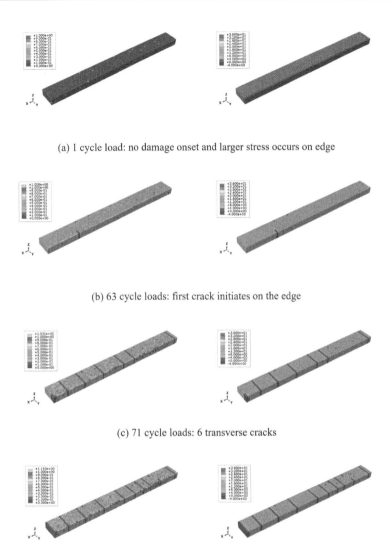

(a) 1 cycle load: no damage onset and larger stress occurs on edge

(b) 63 cycle loads: first crack initiates on the edge

(c) 71 cycle loads: 6 transverse cracks

(d) 73 cycle loads: 9 transverse cracks

FIGURE 6.8   Matrix damage evolution of the 90° layer when the strain boundary condition is 0.6%: left column is the damage onset index, right column is stress $\sigma_{22} + \sigma_{33}$. (a) 1 cycle load: No damage onset and larger stress occurs on edge; (b) 63 cycle loads: First crack initiates on the edge; (c) 71 cycle loads: 6 transverse cracks; (d) 73 cycle loads: 9 transverse cracks.

that a displacement boundary condition is implemented on surface ABFE, as shown in Figure 6.7(a). The impact of stress levels on the results can be incorporated by adjusting the displacement conditions.

First, the influence of the cyclic load number on transverse cracking behavior is investigated. Figure 6.8 illustrates the distribution of the damage onset index and stress ($\sigma22 + \sigma33$) in the 90° layer following a certain number of cyclic loads, with the applied strain at 0.6% (i.e., Ux = 0.6 mm) and a fixed load frequency of 5 Hz. The findings reveal that the damage onset index is less than 1, and the stress $\sigma22+\sigma33$ is higher at the edges than in the interior during the first cycle of the load. Considering that $\sigma22+\sigma33$ is the dominant factor for matrix tensile failure mode, this study primarily focuses on this value. After 63 cycles, the damage onset index reaches 1 at the edge, and stress $\sigma22+\sigma33$ in the damaged area decreases. As the cyclic load increases to 71 and 73, the number of cracks becomes 6 and 9, respectively, supporting the experimental observation that the number of transverse cracks increases with cyclic load.

Figure 6.9(a) demonstrates the impact of stress level on transverse cracking behavior, where the applied strain increases from 0.7% to 0.8%. It is noteworthy that initial transverse cracks form after 38 cycles when the strain boundary condition is 0.7% and after 23 when it is 0.8%. This supports the assertion that a higher stress level leads to earlier failure. Moreover, the transverse crack density rises rapidly after the initial crack formation and then increases slowly, aligning with experimental results.

The influence of load frequency is also examined in Figure 6.9(b). Unlike the effect of stress level, as load frequency increases, the initiation of initial transverse cracks is delayed. Initial cracks form after 13 cycles when the frequency is 2.5 Hz, 23 at 5 Hz, and 44 at 10 Hz. Interestingly, the effect of load frequency on transverse cracking behavior cannot be addressed using traditional empirical formulations such as Paris' law, which is based on the energy release rate. However, the proposed entropy-based failure criterion effectively addresses this. Thus, the proposed model offers a more comprehensive understanding of the effect of load frequency on transverse crack growth behaviors in CFRPs than conventional methods.

6.5 SUMMARY

The study has explored the transverse cracking behavior of CFRP ply under cyclic load using experimental and numerical methods. The key findings include the following points.

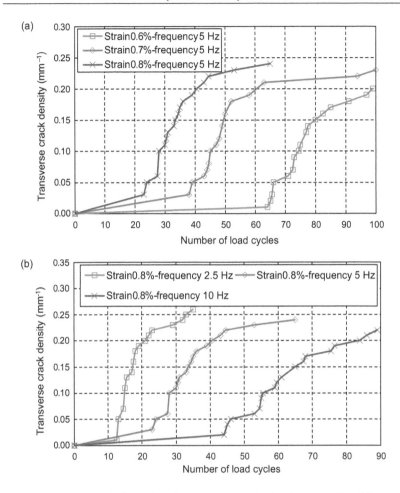

FIGURE 6.9 Numerical results of transverse crack density regarding (a) stress level dependence and (b) frequency dependence.

Experimental results demonstrated that cyclic load level and frequency significantly impact crack accumulation behavior. Specifically, two clear transverse cracks were observed after 10^4 cyclic loads, with 37 transverse cracks forming after 10^5 cycles. When the load frequency increased from 5 Hz to 10 Hz, the final count of transverse cracks decreased from 29 to 11.

In an effort to predict the long-term lifespan of CFRP laminate under fatigue loads, the study introduced an entropy-based failure criterion. This was used to simulate progressive damage and transverse cracking behavior in

CFRP ply. Numerical results showed that as stress levels increased, the formation of initial transverse cracks occurred earlier. Conversely, as load frequency increased, the formation of initial transverse cracks was delayed. For example, when the load frequency was set at 5 Hz, initial transverse cracks formed after 63 cyclic loads at a strain boundary condition of 0.6%. When the strain boundary condition was at 0.7% and 0.8%, the cracks formed after 38 and 23 cyclic loads, respectively.

In addition, when the load frequency was raised from 2.5 Hz to 10 Hz, the number of cyclic loads leading to the initial crack formation increased from 13 to 44. A comparison of the proposed failure model to reference results confirms its ability to account for the effects of cyclic load number, stress level, and load frequency on transverse cracking behavior.

A significant contribution of this study is the proposed entropy-based failure criterion, which effectively models the effect of load frequency on transverse cracking behavior, an aspect not addressed by Paris' law. However, further research is necessary for accurately predicting the lifespan of CFRPs under cyclic loading, such as work on more efficient computational frameworks and the examination of heat generation under cyclic loading. These areas will be the focus of future studies.

REFERENCES

[1] W. Qi, W. Yao, H. Shen, A bi-directional damage model for matrix cracking evolution in composite laminates under fatigue loadings, International Journal of Fatigue 134 (2020).

[2] S. Tamboura, M.A. Laribi, J. Fitoussi, M. Shirinbayan, R.T. Bi, A. Tcharkhtchi, H.B. Dali, Damage and fatigue life prediction of short fiber reinforced composites submitted to variable temperature loading: Application to Sheet Molding Compound composites, International Journal of Fatigue 138 (2020).

[3] R. Aoki, R. Higuchi, T. Yokozeki, Fatigue simulation for progressive damage in CFRP laminates using intra-laminar and inter-laminar fatigue damage models, International Journal of Fatigue 143 (2021).

[4] J.J. Xiong, Y.T. Zhu, C.Y. Luo, Y.S. Li, Fatigue-driven failure criterion for progressive damage modelling and fatigue life prediction of composite structures, International Journal of Fatigue 145 (2021).

[5] A.P. Vassilopoulos, The history of fiber-reinforced polymer composite laminate fatigue, International Journal of Fatigue 134 (2020).

[6] D. Di Maio, G. Voudouris, I.A. Sever, Investigation of fatigue damage growth and self-heating behaviour of cross-ply laminates using simulation-driven dynamic test, International Journal of Fatigue 155 (2022).

[7] J. Guo, W. Wen, H. Zhang, H. Cui, A mesoscale fatigue progressive damage model for 3D woven composites, International Journal of Fatigue 152 (2021).

[8] M.A. Laribi, S. Tamboura, J. Fitoussi, M. Shirinbayan, R.T. Bi, A. Tcharkhtchi, H.B. Dali, Microstructure dependent fatigue life prediction for short fibers reinforced composites: Application to sheet molding compounds, International Journal of Fatigue 138 (2020).

[9] A.V. Movahedi-Rad, G. Eslami, T. Keller, A novel fatigue life prediction methodology based on energy dissipation in viscoelastic materials, International Journal of Fatigue 152 (2021).

[10] M. Brod, A. Dean, R. Rolfes, Numerical life prediction of unidirectional fiber composites under block loading conditions using a progressive fatigue damage model, International Journal of Fatigue 147 (2021).

[11] P. Gholami, M.A. Farsi, M.A. Kouchakzadeh, Stochastic fatigue life prediction of Fiber-Reinforced laminated composites by continuum damage Mechanics-based damage plastic model, International Journal of Fatigue 152 (2021).

[12] N.M. Chowdhury, R. Healey, J. Wang, W.K. Chiu, C. Wallbrink, Using a residual strength model to predict mode II delamination failure of composite materials under block fatigue loading, International Journal of Fatigue 135 (2020).

[13] J. Song, W. Wen, H. Cui, L. Li, Weft direction tension-tension fatigue responses of layer-to-layer 3D angle-interlock woven composites at room and elevated temperatures, International Journal of Fatigue 139 (2020).

[14] J. Koyanagi, S. Yoneyama, A. Nemoto, J.D.D. Melo, Time and temperature dependence of carbon/epoxy interface strength, Composites Science and Technology 70(9) (2010) 1395–1400.

[15] J. Koyanagi, S. Ogihara, H. Nakatani, T. Okabe, S. Yoneyama, Mechanical properties of fiber/matrix interface in polymer matrix composites, Advanced Composite Materials 23 (2014) 551–570.

[16] J. Koyanagi, M. Nakada, Y. Miyano, Tensile strength at elevated temperature and its applicability as an accelerated testing methodology for unidirectional composites, Mechanics of Time-Dependent Materials 16(1) (2012) 19–30.

[17] M. Sato, S. Shirai, J. Koyanagi, Y. Ishida, Y. Kogo, Numerical simulation for strain rate and temperature dependence of transverse tensile failure of unidirectional carbon fiber-reinforced plastics, Journal of Composite Materials 53(28–30) (2019) 4305–4312.

[18] Y. Sawamura, Y. Yamazaki, S. Yoneyama, J. Koyanagi, Multi-scale numerical simulation of impact failure for cylindrical CFRP, Advanced Composite Materials 30 (2021) 19–38.

[19] J. Koyanagi, Y. Sato, T. Sasayama, T. Okabe, S. Yoneyama, Numerical simulation of strain-rate dependent transition of transverse tensile failure mode in fiber-reinforced composites, Composites Part A: Applied Science and Manufacturing 56 (2014) 136–142.

[20] L.P. Canal, J. Segurado, J. Llorca, Failure surface of epoxy-modified fiber-reinforced composites under transverse tension and out-of-plane shear, International Journal of Solids and Structures 46(11–12) (2009) 2265–2274.

[21] Y. Kumagai, S. Onodera, M. Salviato, T. Okabe, Multiscale analysis and experimental validation of crack initiation in quasi-isotropic laminates, International Journal of Solids and Structures 193–194 (2020) 172–191.

[22] O. Pierard, J. Llorca, J. Segurado, I. Doghri, Micromechanics of particle-reinforced elasto -viscoplastic composites: Finite element simulations versus affine homogenization, International Journal of Plasticity 23(6) (2007) 1041–1060.

[23] M. Sato, K. Hasegawa, J. Koyanagi, R. Higuchi, Y. Ishida, Residual strength prediction for unidirectional CFRP using a nonlinear viscoelastic constitutive equation considering entropy damage, Composites Part A: Applied Science and Manufacturing 141 (2021).

[24] N. Takase, J. Koyanagi, K. Mori, T. Sakai, Molecular dynamics simulation for evaluating fracture entropy of a polymer material under various combined stress states, Materials 14(8) (2021).

[25] B. Mohammadi, M.M. Shokrieh, M. Jamali, A. Mahmoudi, B. Fazlali, Damage-entropy model for fatigue life evaluation of off-axis unidirectional composites, Composite Structures 270 (2021).

[26] B. Mohammadi, A. Mahmoudi, Developing a new model to predict the fatigue life of cross-ply laminates using coupled CDM-entropy generation approach, Theoretical and Applied Fracture Mechanics 95 (2018) 18–27.

[27] J. Huang, H. Yang, W. Liu, K. Zhang, A. Huang, Confidence level and reliability analysis of the fatigue life of CFRP laminates predicted based on fracture fatigue entropy, International Journal of Fatigue 156 (2022).

[28] T. Sakai, N. Takase, Y. Oya, J. Koyanagi, A possibility for quantitative detection of mechanically-induced invisible damage by thermal property measurement via entropy generation for a polymer material, Materials 15(3) (2022) 737.

[29] J. Koyanagi, A. Mochizuki, R. Higuchi, V.B.C. Tan, T.E. Tay, Finite element model for simulating entropy-based strength-degradation of carbon-fiber-reinforced plastics subjected to cyclic loadings, International Journal of Fatigue 165 (2022).

[30] J. Koyanagi, Comparison of a viscoelastic frictional interface theory with a kinetic crack growth theory in unidirectional composites, Composites Science and Technology 69(13) (2009) 2158–2162.

[31] K. Fukui, T. Tadachi, H. Yamashita, J. Koyanagi, S. Ogihara, Finite element analysis of edge gripper processing of carbon fiber composite cables, Advanced Composite Materials 31 (2022) 583–599.

[32] H. Deng, A. Mochizuki, M. Fikry, S. Abe, S. Ogihara, J. Koyanagi, Numerical and experimental studies for fatigue damage accumulation of CFRP cross-ply laminates based on entropy failure criterion, Materials 16(1) (2023).

Future Prospectives

7

This book has addressed the durability of CFRPs, focusing on continuous fiber CFRP and discussing it based on micromechanics. Specifically, we have examined the strength and durability in the fiber direction and transverse direction. Durability aspects covered in this book include CSR strength, creep strength, and fatigue strength. Numerous comparisons with experimental data have been presented to demonstrate the validity of durability predictions based on micromechanics. As a result, we have established a foundation for understanding the durability of CFRP.

Looking ahead, the following endeavors are anticipated:

(1) Prediction of interlayer debonding (delamination) in CFRP laminates.

The prediction of interlayer debonding (delamination) in CFRP laminates is a significant focus of future research. While this book covers numerical simulations at the nano, micro, and mesoscales, there is still a need to establish a quantitative link between these different scales. Once this link is established, the next step would involve addressing delamination in CFRP laminates, which occurs at the millimeter scale. Delamination is considered a critical form of damage in CFRP laminates, and simulating its occurrence is of paramount importance. Cohesive zone modeling (CZM) is commonly employed to simulate delamination. However, it is important to note that CZM typically does not incorporate time dependence or cyclic loading dependence, as delamination is often triggered by transverse cracking. Therefore, by incorporating time dependence and cyclic loading effects on transverse cracking, it is expected that the delamination process can also be accounted for in a time-dependent manner. By further investigating and understanding the mechanisms that lead to delamination, researchers can enhance the predictive capabilities of numerical simulations and ultimately improve the durability of CFRP laminates. The successful simulation of delamination would be a significant advancement in the field, providing valuable insights for optimizing the design and performance of composite structures.

(2) Evaluation of durability in discontinuous fiber CFRP.

In addition to the focus on continuum fiber reinforcement in aerospace applications, this book acknowledges the significance of short fiber-reinforced plastics (SFRP) in automobile applications. SFRP offers advantages in terms of fabrication and assembly, making it a desirable choice. However, the long-term durability of SFRP remains a crucial factor to consider in these applications. Currently, there are limited articles that numerically simulate the long-term durability of SFRP. The complex nature of the numerical simulation, especially for static failure, poses challenges. Nevertheless, by employing an entropy-based failure model, it becomes possible to simulate the time-dependent failure of SFRP using finite element analysis. One key aspect in this analysis is the consideration of friction between debonded fibers and the matrix. Friction allows for an increase in fiber stress, preventing complete debonding and maintaining load-bearing capacity. Without friction, the debonding process would continue until all fibers separate from the matrix, resulting in significantly reduced material strength. By incorporating time-dependent friction, it is believed that the long-term durability of SFRP can be successfully simulated. This aspect presents an intriguing avenue for future research, as it has the potential to advance our understanding and prediction of the long-term durability of SFRP materials. By addressing the challenges associated with numerical simulations of SFRP, we can further enhance the performance and reliability of these materials in automotive applications.

(3) Durability under random loading conditions.

In practical applications, composite materials are subjected to various loading conditions. However, it is often challenging to accurately predict the durability of composites under real-world random loading scenarios based solely on results obtained from simplified loading conditions such as constant stress ratio (CSR), creep, or monotonic fatigue. In such cases, two options can be considered: linear cumulative damage (LCD) law and the kinetic crack growth theory proposed by Christensen. However, both LCD law and the kinetic crack growth theory are macroscopic theories, and their applicability to heterogeneous composite materials remains uncertain. On the other hand, the entropy-based failure criterion provides a more versatile and comprehensive approach. By utilizing the entropy-based failure criterion, it becomes possible to predict and simulate the residual strength, remaining lifetime, damage state, and degradation state of composite materials under random loading conditions. The entropy-based failure criterion offers a universal applicability for predicting the actual durability of composite materials. It allows for a more accurate assessment of the material's response to random loading scenarios, taking into account the complex behavior and heterogeneity of composites. By employing this criterion, researchers and engineers can make more informed decisions regarding the design and durability assessment of composite materials in real-world applications.

(4) Development of simplified models.

As we strive for more accurate predictions of nonlinear behavior, additional terms are often incorporated into constitutive equations. However, this can lead to increased model complexity, especially when the predictions do not fully align with experimental results. Consequently, the number of constants in the model grows, making the parameter of identification challenging and limiting its applicability. In order to make durability predictions more accessible and practical, it is crucial to simplify the models. The simpler the model, the better its applicability becomes for a wider range of users. In the near future, it is important to shift our focus from expanding models to shrinking them, enhancing their practicality. This involves exploring simpler approaches such as LCD and kinetic crack growth theory, provided their validity can be verified. The pursuit of simplifying various models is an intriguing avenue for future research. By developing more streamlined and accessible models, we can improve the applicability and usefulness of durability predictions in a broader range of practical applications.

(5) Design considerations at the molecular level.

The entropy damage criterion suggests that a desirable failure scenario involves higher failure entropy, lower Arrhenius activation energy, and reduced viscoelasticity. To develop such resin materials, an approach from the molecular level becomes crucial. Molecular dynamics simulations provide a valuable tool for exploring and designing new polymer materials. Additionally, the integration of neural networks, a form of artificial intelligence, with molecular dynamics simulations offers great potential for material development. This combination allows for the customization of materials, including the creation of vitrimer materials and biodegradable polymers, through computer simulations. We are entering an era where computer simulations enable the design and development of tailored materials, revolutionizing the field of material science.

AT THE END

This book endeavors to illuminate the intricate world of carbon-fiber-reinforced plastics (CFRP) and their durability. Through a comprehensive journey spanning molecular dynamics simulations to macroscale predictions, we have unraveled the mysteries of composite behavior under varying conditions. Our exploration into micromechanics, viscoelasticity, and damage mechanisms has paved the way for a deeper understanding of CFRP's response to time, temperature, and cyclic loading. The introduction of entropy-based principles has offered a novel lens through which we can predict and quantify degradation, remaining lifetime, and residual strength with unparalleled accuracy. As we stand at the crossroads of innovation, this work serves as a cornerstone, beckoning researchers and engineers to continue pushing the boundaries of composite material science, design, and application.

Index

Note: Page numbers in *italics* indicate a figure and page numbers in **bold** indicate a table on the corresponding page.